Brauer Groups and the Cohomology of Graded Rings

PURE AND APPLIED MATHEMATICS

A Program of Monographs, Textbooks, and Lecture Notes

MONOGRAPHS AND TEXTBOOKS IN PURE AND APPLIED MATHEMATICS

1. *K. Yano,* Integral Formulas in Riemannian Geometry (1970)*(out of print)*
2. *S. Kobayashi,* Hyperbolic Manifolds and Holomorphic Mappings (1970) *(out of print)*
3. *V. S. Vladimirov,* Equations of Mathematical Physics (A. Jeffrey, editor; A. Littlewood, translator) (1970) *(out of print)*
4. *B. N. Pshenichnyi,* Necessary Conditions for an Extremum (L. Neustadt, translation editor; K. Makowski, translator) (1971)
5. *L. Narici, E. Beckenstein, and G. Bachman,* Functional Analysis and Valuation Theory (1971)
6. *S. S. Passman,* Infinite Group Rings (1971)
7. *L. Dornhoff,* Group Representation Theory (in two parts). Part A: Ordinary Representation Theory. Part B: Modular Representation Theory (1971, 1972)
8. *W. Boothby and G. L. Weiss (eds.),* Symmetric Spaces: Short Courses Presented at Washington University (1972)
9. *Y. Matsushima,* Differentiable Manifolds (E. T. Kobayashi, translator) (1972)
10. *L. E. Ward, Jr.,* Topology: An Outline for a First Course (1972) *(out of print)*
11. *A. Babakhanian,* Cohomological Methods in Group Theory (1972)
12. *R. Gilmer,* Multiplicative Ideal Theory (1972)
13. *J. Yeh,* Stochastic Processes and the Wiener Integral (1973) *(out of print)*
14. *J. Barros-Neto,* Introduction to the Theory of Distributions (1973) *(out of print)*
15. *R. Larsen,* Functional Analysis: An Introduction (1973) *(out of print)*
16. *K. Yano and S. Ishihara,* Tangent and Cotangent Bundles: Differential Geometry (1973) *(out of print)*
17. *C. Procesi,* Rings with Polynomial Identities (1973)
18. *R. Hermann,* Geometry, Physics, and Systems (1973)
19. *N. R. Wallach,* Harmonic Analysis on Homogeneous Spaces (1973) *(out of print)*
20. *J. Dieudonné,* Introduction to the Theory of Formal Groups (1973)
21. *I. Vaisman,* Cohomology and Differential Forms (1973)
22. *B.-Y. Chen,* Geometry of Submanifolds (1973)
23. *M. Marcus,* Finite Dimensional Multilinear Algebra (in two parts) (1973, 1975)
24. *R. Larsen,* Banach Algebras: An Introduction (1973)
25. *R. O. Kujala and A. L. Vitter (eds.),* Value Distribution Theory: Part A; Part B: Deficit and Bezout Estimates by Wilhelm Stoll (1973)
26. *K. B. Stolarsky,* Algebraic Numbers and Diophantine Approximation (1974)
27. *A. R. Magid,* The Separable Galois Theory of Commutative Rings (1974)
28. *B. R. McDonald,* Finite Rings with Identity (1974)
29. *J. Satake,* Linear Algebra (S. Koh, T. A. Akiba, and S. Ihara, translators) (1975)

30. *J. S. Golan,* Localization of Noncommutative Rings (1975)
31. *G. Klambauer,* Mathematical Analysis (1975)
32. *M. K. Agoston,* Algebraic Topology: A First Course (1976)
33. *K. R. Goodearl,* Ring Theory: Nonsingular Rings and Modules (1976)
34. *L. E. Mansfield,* Linear Algebra with Geometric Applications: Selected Topics (1976)
35. *N. J. Pullman,* Matrix Theory and Its Applications (1976)
36. *B. R. McDonald,* Geometric Algebra Over Local Rings (1976)
37. *C. W. Groetsch,* Generalized Inverses of Linear Operators: Representation and Approximation (1977)
38. *J. E. Kuczkowski and J. L. Gersting,* Abstract Algebra: A First Look (1977)
39. *C. O. Christenson and W. L. Voxman,* Aspects of Topology (1977)
40. *M. Nagata,* Field Theory (1977)
41. *R. L. Long,* Algebraic Number Theory (1977)
42. *W. F. Pfeffer,* Integrals and Measures (1977)
43. *R. L. Wheeden and A. Zygmund,* Measure and Integral: An Introduction to Real Analysis (1977)
44. *J. H. Curtiss,* Introduction to Functions of a Complex Variable (1978)
45. *K. Hrbacek and T. Jech,* Introduction to Set Theory (1978)
46. *W. S. Massey,* Homology and Cohomology Theory (1978)
47. *M. Marcus,* Introduction to Modern Algebra (1978)
48. *E. C. Young,* Vector and Tensor Analysis (1978)
49. *S. B. Nadler, Jr.,* Hyperspaces of Sets (1978)
50. *S. K. Segal,* Topics in Group Rings (1978)
51. *A. C. M. van Rooij,* Non-Archimedean Functional Analysis (1978)
52. *L. Corwin and R. Szczarba,* Calculus in Vector Spaces (1979)
53. *C. Sadosky,* Interpolation of Operators and Singular Integrals: An Introduction to Harmonic Analysis (1979)
54. *J. Cronin,* Differential Equations: Introduction and Quantitative Theory (1980)
55. *C. W. Groetsch,* Elements of Applicable Functional Analysis (1980)
56. *I. Vaisman,* Foundations of Three-Dimensional Euclidean Geometry (1980)
57. *H. I. Freedman,* Deterministic Mathematical Models in Population Ecology (1980)
58. *S. B. Chae,* Lebesgue Integration (1980)
59. *C. S. Rees, S. M. Shah, and C. V. Stanojević,* Theory and Applications of Fourier Analysis (1981)
60. *L. Nachbin,* Introduction to Functional Analysis: Banach Spaces and Differential Calculus (R. M. Aron, translator) (1981)
61. *G. Orzech and M. Orzech,* Plane Algebraic Curves: An Introduction Via Valuations (1981)
62. *R. Johnsonbaugh and W. E. Pfaffenberger,* Foundations of Mathematical Analysis (1981)
63. *W. L. Voxman and R. H. Goetschel,* Advanced Calculus: An Introduction to Modern Analysis (1981)
64. *L. J. Corwin and R. H. Szcarba,* Multivariable Calculus (1982)
65. *V. I. Istrătescu,* Introduction to Linear Operator Theory (1981)
66. *R. D. Järvinen,* Finite and Infinite Dimensional Linear Spaces: A Comparative Study in Algebraic and Analytic Settings (1981)

67. *J. K. Beem and P. E. Ehrlich,* Global Lorentzian Geometry (1981)
68. *D. L. Armacost,* The Structure of Locally Compact Abelian Groups (1981)
69. *J. W. Brewer and M. K. Smith, eds.,* Emmy Noether: A Tribute to Her Life and Work (1981)
70. *K. H. Kim,* Boolean Matrix Theory and Applications (1982)
71. *T. W. Wieting,* The Mathematical Theory of Chromatic Plane Ornaments (1982)
72. *D. B. Gauld,* Differential Topology: An Introduction (1982)
73. *R. L. Faber,* Foundations of Euclidean and Non-Euclidean Geometry (1983)
74. *M. Carmeli,* Statistical Theory and Random Matrices (1983)
75. *J. H. Carruth, J. A. Hildebrant, and R. J. Koch,* The Theory of Topological Semigroups (1983)
76. *R. L. Faber,* Differential Geometry and Relativity Theory: An Introduction (1983)
77. *S. Barnett,* Polynomials and Linear Control Systems (1983)
78. *G. Karpilovsky,* Commutative Group Algebras (1983)
79. *F. Van Oystaeyen and A. Verschoren,* Relative Invariants of Rings: The Commutative Theory (1983)
80. *I. Vaisman,* A First Course in Differential Geometry (1984)
81. *G. W. Swan,* Applications of Optimal Control Theory in Biomedicine (1984)
82. *T. Petrie and J. D. Randall,* Transformation Groups on Manifolds (1984)
83. *K. Goebel and S. Reich,* Uniform Convexity, Hyperbolic Geometry, and Nonexpansive Mappings (1984)
84. *T. Albu and C. Năstăsescu,* Relative Finiteness in Module Theory (1984)
85. *K. Hrbacek and T. Jech,* Introduction to Set Theory, Second Edition, Revised and Expanded (1984)
86. *F. Van Oystaeyen and A. Verschoren,* Relative Invariants of Rings: The Noncommutative Theory (1984)
87. *B. R. McDonald,* Linear Algebra Over Commutative Rings (1984)
88. *M. Namba,* Geometry of Projective Algebraic Curves (1984)
89. *G. F. Webb,* Theory of Nonlinear Age-Dependent Population Dynamics (1985)
90. *M. R. Bremner, R. V. Moody, and J. Patera,* Tables of Dominant Weight Multiplicities for Representations of Simple Lie Algebras (1985)
91. *A. E. Fekete,* Real Linear Algebra (1985)
92. *S. B. Chae,* Holomorphy and Calculus in Normed Spaces (1985)
93. *A. J. Jerri,* Introduction to Integral Equations with Applications (1985)
94. *G. Karpilovsky,* Projective Representations of Finite Groups (1985)
95. *L. Narici and E. Beckenstein,* Topological Vector Spaces (1985)
96. *J. Weeks,* The Shape of Space: How to Visualize Surfaces and Three-Dimensional Manifolds (1985)
97. *P. R. Gribik and K. O. Kortanek,* Extremal Methods of Operations Research (1985)
98. *J.-A. Chao and W. A. Woyczynski, eds.,* Probability Theory and Harmonic Analysis (1986)
99. *G. D. Crown, M. H. Fenrick, and R. J. Valenza,* Abstract Algebra (1986)
100. *J. H. Carruth, J. A. Hildebrant, and R. J. Koch,* The Theory of Topological Semigroups, Volume 2 (1986)

101. *R. S. Doran and V. A. Belfi,* Characterizations of C*-Algebras: The Gelfand-Naimark Theorems (1986)
102. *M. W. Jeter,* Mathematical Programming: An Introduction to Optimization (1986)
103. *M. Altman*, A Unified Theory of Nonlinear Operator and Evolution Equations with Applications: A New Approach to Nonlinear Partial Differential Equations (1986)
104. *A. Verschoren*, Relative Invariants of Sheaves (1987)
105. *R. A. Usmani,* Applied Linear Algebra (1987)
106. *P. Blass and J. Lang,* Zariski Surfaces and Differential Equations in Characteristic $p > 0$ (1987)
107. *J. A. Reneke, R. E. Fennell, and R. B. Minton.* Structured Hereditary Systems (1987)
108. *H. Busemann and B. B. Phadke*, Spaces with Distinguished Geodesics (1987)
109. *R. Harte*, Invertibility and Singularity for Bounded Linear Operators (1988).
110. *G. S. Ladde, V. Lakshmikantham, and B. G. Zhang*, Oscillation Theory of Differential Equations with Deviating Arguments (1987)
111. *L. Dudkin, I. Rabinovich, and I. Vakhutinsky*, Iterative Aggregation Theory: Mathematical Methods of Coordinating Detailed and Aggregate Problems in Large Control Systems (1987)
112. *T. Okubo*, Differential Geometry (1987)
113. *D. L. Stancl and M. L. Stancl*, Real Analysis with Point-Set Topology (1987)
114. *T. C. Gard*, Introduction to Stochastic Differential Equations (1988)
115. *S. S. Abhyankar*, Enumerative Combinatorics of Young Tableaux (1988)
116. *H. Strade and R. Farnsteiner*, Modular Lie Algebras and Their Representations (1988)
117. *J. A. Huckaba*, Commutative Rings with Zero Divisors (1988)
118. *W. D. Wallis*, Combinatorial Designs (1988)
119. *W. Więsław,* Topological Fields (1988)
120. *G. Karpilovsky*, Field Theory: Classical Foundations and Multiplicative Groups (1988)
121. *S. Caenepeel and F. Van Oystaeyen,* Brauer Groups and the Cohomology of Graded Rings (1989)
122. *W. Kozlowski,* Modular Function Spaces (1988)

Other Volumes in Preparation

Brauer Groups and the Cohomology of Graded Rings

STEFAAN CAENEPEEL
Free University of Brussels, VUB
Brussels, Belgium

FREDDY VAN OYSTAEYEN
University of Antwerp, UIA
Wilrijk, Belgium

MARCEL DEKKER, INC. New York and Basel

Library of Congress Cataloging-in-Publication Data

Caenepeel, S. (Stefaan), [date]
Brauer groups and the cohomology of graded rings / S. Caenepeel, F. van Oystaeyen.
p. cm. — (Monographs and textbooks in pure and applied mathematics ; v. 121)
Includes index.
ISBN 0-8247-7978-9
1. Azumaya algebras. 2. Brauer groups. 3. Graded rings. 4. Homology theory. I. Oystaeyen, F. van, [date]. II. Title. III. Series.
QA251.5.C34 1988
512′.24—dc19 88-13112
CIP

MARCEL DEKKER, INC.
270 Madison Avenue, New York, New York 10016

Current printing (last digit):
10 9 8 7 6 5 4 3 2 1

PRINTED IN THE UNITED STATES OF AMERICA

Preface

The aim of this work is to study Azumaya algebras over **Z**-graded commutative rings and to research the effect, on classical methods and results, of the existence of the **Z**-gradation on the ground ring, as well as to derive information about classical (ungraded) objects defined over the graded ring making use of graded data. Therefore, it is necessary to introduce various notions defined intrinsically in graded terms extending the notions most frequently used as basic ingredients in the theory of Azumaya algebras: projective modules, Morita theory, separability and Galois extensions of commutative rings, crossed products and Galois cohomology, Picard groups, and the Brauer group. Concerning this theory, one's first impression might be that it is sufficient to replace the category of (left) modules by that of (left) graded modules, but at certain points this routine quickly fails; for example, the graded Picard group of a gr-semilocal ring need not be trivial.

If we are to investigate the effect of the existence of a **Z**-gradation on the ground ring on all methods and techniques mentioned above, the first prerequisite is that we be able to recognize which Azumaya algebras over the graded ring bear the information of graded nature. Here it is obvious that those Azumaya algebras allowing a **Z**-gradation extending the gradation of the center are the first candidates for a detailed study. Restricting attention to such Azumaya algebras leads us to develop a theory of graded Brauer groups formulated almost completely intrinsically in terms of graded rings and graded modules. In certain respects this theory has some similarity with the theory of the Brauer–Wall group or the Brauer–Long group (we come back to this aspect in Chapter VI), the essential difference being that our "graded Azumaya algebras" are common Azumaya algebras when considered as ungraded objects, but are equipped with a graded structure extending that of the center. This also makes it clear that the so-called graded Brauer group must be linked to the usual Brauer group, by applying the forgetful functor to the graded structure. Let us point out that a graded module P over the **Z**-graded ring R is projective in the category of graded R-modules (denoted by R-gr) if and only if it is projective as an R-module, but on the other hand, if $\mathrm{End}_R(Q)$ is a **Z**-graded R-algebra for some projective R-module Q, then Q need not necessarily be **Z**-graded (counterexamples exist). Therefore the forgetful functor gives rise to a group morphism $\mathrm{Br}_g(R) \to \mathrm{Br}(R)$, which need not be injective or surjective. In this context the main problem for us is to find how far $\mathrm{Br}_g(R)$ determines $\mathrm{Br}(R)$. This is the classical main problem, for graded rings and modules, of deriving properties and structural results for ungraded objects defined over a graded ring from graded data emanating from the study of suitable graded objects associated to the original object.

Throughout the book, we have strived for a strictly algebraic presentation of the results, even if several connections with algebraic geometry may be detected. We have not included details of a geometrical nature but contented ourselves with pointing out such links whenever they exist. In fact, on one hand, there is a link with the theory of Brauer groups of projective varieties; we discuss this aspect in Chapter VI. On the other hand, the methods used in Chapter V definitely have a more geometrical flavor. As an illustration of this geometrical aspect of the theory we point to the different variations of the notion of a strongly graded ring that appear quite naturally in Chapters IV and V. For the study of K-theory (Grothendieck group) and module theory (cf. Sections III.2 and III.3), it is necessary to introduce the concept "strongly graded with respect to the flat topology," but if one is interested in the Brauer group, however, it is the notion "strongly graded on the étale site" that is needed.

In Chapters I and II we provide some information that is almost purely graded in nature and we include graded equivalents of results and techniques that are well established in the ungraded case, for example, graded Galois extensions, graded completion and Henselization, the join of gr-Henselian rings. One of the striking points is that, in general, we have no graded equivalent to Artin's refinement theorem on the join of Hensel rings; a counterexample is presented, and it is shown that the theorem remains valid if we restrict attention to rings that are strongly graded on the étale site.

In Chapter III we present several "crossed product theorems," and although these are still completely in the graded context, the cohomological methods introduced here will eventually lead, in Chapters IV and V, to the determination of some useful relations between the Brauer group $\mathrm{Br}_g(R)$ and $\mathrm{Br}(R)$, at least for certain classes of graded rings. The crossed product results tie in with some observations concerning strongly graded rings (sometimes called generalized crossed products) that come in handy in stating crossed product structure theorems over ground rings with nontrivial Picard group. In the case of graded Azumaya algebras split by a graded Galois extension, the generalized crossed product structure mentioned above is equivalent to the Chase–Harrison–Rosenberg sequence in Galois cohomology. The class of graded Krull domains plays an important part throughout this theory, because for such rings, $\mathrm{Br}_g(R) \subset \mathrm{Br}(R)$ holds. In fact, if R is a Krull domain which is strongly graded on the étale site, and which is such that n is invertible in R, then ${}_n\mathrm{Br}(R) = {}_n\mathrm{Br}_g(R_{(n)})$, where $R_{(n)}$ is obtained from R by blowing up the gradation on R as follows: $(R_{(n)})_{nm} = R_m$ for all m in $\mathbf{Z}$.

Since not every Azumaya algebra can be split by a Galois extension, not necessarily even by a projective separable extension of the ground ring, it will be necessary to introduce étale cohomology to obtain suitable cohomological interpretations of the (graded) Brauer group. In Chapter V we adapt the étale cohomological methods to the graded situation, and in doing so we provide some new facts about $\mathrm{Br}_g(R)$. For example, it is proved that there is a natural embedding of $\mathrm{Br}_g(R)$ into a graded étale cohomology group $H^2_{\mathrm{gr}}(R_{\text{ét}_g}, U)$. Some technical problems occur if R is not strongly graded on the étale site: first, the graded Picard group Pic_g of a gr-local ring is not trivial, and consequently, the sheaf associated to Pic_g does not vanish; second, we have no graded version of Artin's refinement theorem, which implies that the étale cohomology groups are not described completely by Amitsur or Čech cohomology. We overcome the first problem by looking at a weaker version of the graded Picard group, Pic^g, which embeds in the usual Picard group. The second difficulty can be avoided in two different ways: we may describe the étale cohomology groups using hypercoverings and Verdier's refinement theorem (cf. Section V.2), or we may adapt classical methods developed by Villamayor and Zelinsky to the present situation (cf. Section V.5). At the same time, we obtain cohomological proofs of facts that would otherwise be proved in a cumbersome way; for example, $\mathrm{Br}_g(R)$ is a torsion group. Using some results concerning the Grothendieck group of a strongly graded ring, and again the graded version of Artin's refinement theorem, we obtain a graded equivalent of the Gabber–Hoobler theorem: if R is strongly graded on the étale site, then $\mathrm{Br}_g(R)$ is isomorphic to the torsion part of the étale cohomology group $H^2(R_{\text{ét}_g}, U_0)$.

In the first two sections of Chapter VI, we apply the technique of graded cohomology developed in Chapters III, IV, and V to a fairly different situation. As it happens, a modification of our theory leads to a purely cohomological treatment of the Brauer–Wall group (in the case of a cyclic group of order 2) and the Brauer–Long group (in the case of a finite abelian group). As one may expect, the cohomology of the constant sheaf G appears in the description, and in the case of the Brauer–Long group (where we look at G-dimodule algebras instead of G-graded algebras), the cohomology of the dual sheaf G^* also has to be taken into account. The key point in our discussion is the fact that the sheaf associated to the dimodule version of the Picard group is not zero (in fact, it is $G \times G^*$), and therefore the Brauer–Long group is embedded in a product of cohomology groups instead of just one cohomology group. The multiplication in this product is complicated, but can also be described in cohomological terms. The Brauer–Wall group is described completely, and this provides a shorter proof of some results of C. Small. For the Brauer–Long group, we could obtain a cohomological description of the subgroup consisting of central classes (in the case of a cyclic group), thus providing new and elegant proofs of results of Long, Orzech, Deegan, and Beattie.

The final section deals with the links between the $\mathbf{Z}$-graded Brauer group and the Brauer group of a projective variety. In low dimensions, that is, for curves and surfaces, this theory can be presented without making use of the relative Brauer group introduced in [69]. A consequent approach through the study of rings regular in codimension m provides new short proofs for the main results obtained in [69].

While working on this book we enjoyed stimulating conversations with many people. M. Van den Bergh, R. Hoobler, M.-A. Knus, and M. Beattie contributed some ideas, and we also benefited from the notes of a seminar organized at Lausanne by M. Ojanguren and R. Sridharan. Several people shared a friendly interest in our work: L. Le Bruyn, A. Verschoren, F. Tilborghs, and J. Murre. We are grateful to L. Van Hamme for, while a number theorist, he stimulated the development of pure algebra research at the Free University of Brussels.

Stefaan Caenepeel
Freddy Van Oystaeyen

Contents

PREFACE iii

I. GENERALIZED CROSSED PRODUCTS 1

I.1 Graded Ring Theory 1
I.2 Generalized Crossed Products 3

II. SOME RESULTS ON COMMUTATIVE GRADED RINGS 19

II.1 Arithmetically Graded Rings 19
II.2 Separability and Graded Galois Extensions 34
II.3 Graded Completion and Henselization 46
II.4 The Join of gr-Henselian Rings 61

III. GRADED BRAUER GROUPS AND THE CROSSED PRODUCT THEOREMS 79

III.1 Graded Faithfully Flat Descent 79

III.2 Projective Graded Modules 83
III.3 Grothendieck and Picard Groups of Graded Rings 90
III.4 Brauer Groups of Graded Rings 105
III.5 Graded Cohomology Groups and the Crossed Product Theorem 119

IV. APPLICATION TO SOME SPECIAL CASES 133
IV.1 Brauer Groups of Graded Fields 133
IV.2 Brauer Groups of gr-Local Rings 141
IV.3 The Brauer Group of a Graded Ring Modulo a Graded Ideal 149
IV.4 Brauer Groups of Regular Graded Rings 151

V. ETALE COHOMOLOGY FOR GRADED RINGS 157
V.1 Cohomology on the gr-Etale Site 157
V.2 Hypercoverings and Verdier's Refinement Theorem 166
V.3 Application to the Graded Brauer Group 172
V.4 A Graded Version of Gabber's Theorem 184
V.5 The Villamayor–Zelinsky Approach 190

VI. APPLICATIONS 201
VI.1 The Brauer–Long Group 201
VI.2 The Brauer–Wall Group 228
VI.3 Graded Brauer Groups in a Geometrical Context 235

REFERENCES 251
INDEX 259

Brauer Groups and the Cohomology of Graded Rings

Chapter I

Generalized Crossed Products

I.1 GRADED RING THEORY

For a general study of graded rings and modules, the reader is referred to the monographs [48] and [49] by C. Năstăsescu and F. Van Oystaeyen. The major part of this book is restricted to the study of **Z**-graded rings; only in Sections I.2 and II.1 do we look at G-graded rings.

Let us recall briefly some definitions and elementary properties. A *graded ring of type* G, where G is an arbitrary group, is a ring R, together with a family of additive subgroups $\{R_\sigma : \sigma \in G\}$, such that $R = \oplus_{\sigma \in G} R_\sigma$ and $R_\sigma R_\tau \subset R_{\sigma\tau}$ for all σ, τ in G. If $R_\sigma R_\tau = R_{\sigma\tau}$ for all σ, τ, then R is said to be *strongly graded.* In a similar way, we define *graded (left or right) modules* over R as R-modules together with a family of R_e-submodules such that $M = \oplus_{\sigma \in G} M_\sigma$ and $R_\sigma M_\tau \subset M_{\sigma\tau}$. For a in M_σ, we denote $\deg_M a = \sigma$, and refer to a as being *homogeneous* of degree σ.

The multiplicative set of all homogeneous elements in R or M is denoted $h(R)$ or $h(M)$. An R-linear map $f : M \to N$ is said to be a *morphism of degree* σ if $f(M_\tau) \subset N_{\tau\sigma}$, for all $\tau \in G$.

Let us consider the case $G = \mathbf{Z}$. A morphism of degree 0 is called a *graded morphism.* The set of all morphisms of degree p is denoted by $\mathrm{HOM}_R(M,N)_p$. We define $\mathrm{HOM}_R(M,N)$ to be the graded R-algebra $\oplus_{p\in Z}\mathrm{HOM}_R(M,N)_p$. It may be shown that, in case M is finitely generated, $\mathrm{HOM}_R(M,N) \cong \mathrm{Hom}_R(M,N)$ as ungraded objects. Two graded R-modules M and N are said to be *graded isomorphic* (denoted $M \cong_g N$) if there exists a graded isomorphism $f : M \to N$.

Let R-gr be the category of graded left R-modules and graded R-morphisms. Then $\mathrm{Hom}_{R\text{-gr}}(M,N) = \mathrm{HOM}_R(M,N)_0$, and it can be shown that R-gr is a Grothendieck category having enough projective and injective objects. We remark that there exists a natural functor $U : R\text{-gr} \to R\text{-mod} : M \to U(M) = \underline{M}$, obtained by forgetting the gradations. Also, it is a classical result that R is strongly graded if and only if the categories R-gr and R_0-mod are naturally equivalent.

A graded (or homogeneous) ideal of R (i.e., a graded submodule of R) is called *gr-prime* if $h(R\backslash p)$ is multiplicatively closed, or, equivalently, if it is a graded prime ideal. I is called *gr-maximal* if it is a maximal element in the set of graded ideals. Defining the *graded nilradical* $N^g(R)$ and the *graded Jacobson radical* $J^g(R)$ as the intersections of all gr-prime and gr-maximal ideals, respectively, we find that $N(R) = N^g(R) \subset J(R) \subset J^g(R)$. Also, we have a graded version of Nakayama's lemma: if $M \in R$-gr is finitely generated, then $J^g(R)M \neq M$.

A graded ring R is called a *graded field*, or, more correctly, a *gr-field*, if all nonzero homogeneous elements of R are invertible. It is well known that, in this case, either $R = R_0$ is a trivially graded field, or $R = R_0[X, X^{-1}, \varphi]$, where R_0 is a field, φ an automorphism of R_0, and X and a variable of positive degree satisfying

$Xa = \varphi(a)X$. A commutative graded ring is called *gr-local (gr-semilocal)* if it has exactly one (finitely many) gr-maximal ideals. For a graded prime ideal p, we may gr-localize R at $h(R \setminus p)$. The gr-local ring thus obtained is denoted by $Q_p^g(R)$ or R_p^g. In a similar way, we obtain the gr-semilocalization $Q_{p_1 \cdots p_r}^g(R) = R_{p_1 \cdots p_r}$ at a finite number of graded primes by localizing at $\bigcap_{i=1}^r h(R \setminus p_i)$.

For a graded ring R, $\mathrm{Spec}^g(R)$ denotes the subspace of $\mathrm{Spec}(R)$ consisting of all graded primes of R. A base for the Zariski topology on $\mathrm{Spec}^g(R)$ is given by sets of the form $D^g(f) = \{p \in \mathrm{Spec}^g(R) : f \notin p\} = D(f) \cap \mathrm{Spec}^g(R)$.

Since all minimal primes are graded, $\mathrm{Spec}^g(R)$ is a dense subspace of $\mathrm{Spec}(R)$. The set of all maximal elements in $\mathrm{Spec}^g(R)$ is denoted by $\mathrm{Max}^g(R)$.

I.2 GENERALIZED CROSSED PRODUCTS

In this section, crossed products and generalized crossed products are treated as special cases of graded rings. In this setup it is rather essential to allow nonabelian groups as grading groups, because we want to be able to refer to the more classical crossed product results in the theory of the Brauer group of a ring as special cases. One of the features of the latter theory is the Chase–Harrison–Rosenberg seven-term exact sequence in Galois cohomology; here we will recover it in a more general form. The material we present is in large a rearrangement of some results of Y. Miyashita [45], T. Kanzaki [34], E. Dade [20], F. Van Oystaeyen [67], C. Năstăsescu, F. Van Oystaeyen [49, 50].

I.2.1 We consider a subring B of A and write $Z(B)$ and $Z(A)$ for the respective centers. A two-sided B-submodule P of A is said to be *invertible* in A if and only if there exists a two-sided B-submodule Q of A such that $PQ = QP = B$. We will denote the group of all B-automorphisms of A by $\mathrm{Aut}_B(A)$ and we let $I_B(A)$ be the group of all invertible two-sided B-submodules of A.

I.2.2 Proposition With notation as in I.2.1, $I_B(A)$ is an $\mathrm{Aut}_B(A)$-group and $\mathrm{Aut}_B(A)$ is an $I_B(A)$-group.

Proof: Clearly any B-automorphism σ of A maps P in $I_B(A)$ to some $P^\sigma \in I_B(A)$, and this defines an action of $\mathrm{Aut}_B(A)$. If $P \in I_B(A)$, then one easily verifies that there are canonical isomorphisms $A = PA \cong P \otimes_B A$ and $A = AP^{-1} \cong A \otimes_B P^{-1}$. Hence the map $P \otimes_B A \otimes_B P^{-1} \to A$ which sends $x \otimes a \otimes x'$ to xax' is an isomorphism. For any σ in $\mathrm{Aut}_B(A)$ the map $P(\sigma)$ from A to itself which sends $x \otimes a \otimes x'$ to $x \otimes \sigma(a) \otimes x'$ yields a B-automorphism of A. It is easily verified that we obtain an action of $I_B(A)$ on $\mathrm{Aut}_B(A)$ by letting $P \in I_B(A)$ act on $\sigma \in \mathrm{Aut}_B(A)$ as described above. ∎

I.2.3 If $P \in I_B(A)$ then from $1 \in B = PP^{-1}$ it follows that we may select a decomposition $1 = \sum_i a_i a_i'$ with $a_i \in P$ and $a_i' \in P^{-1}$. For $\sigma, \tau \in \mathrm{Aut}_B(A)$ one easily verifies that

$$\left(\sum_i \tau(a_i)\sigma(a_i')\right)\left(\sum_i \sigma(a_i)\tau(a_i')\right) = 1.$$

The element $\sum_i a_i \otimes a_i'$ is central, because it is the image of 1 under the isomorphism $B \cong P \otimes P^{-1}$; hence $\sum_i \tau(a_i)\sigma(a_i') \in C_A(B) = \{x \in A; bx = xb \text{ for all } b \in B\}$ and it is a unit in this ring, with

$$\left(\sum_i \tau(a_i)\sigma(a_i')\right)^{-1} = \sum_i \sigma(a_i)\tau(a_i').$$

If we put $u(\sigma) = \sum_i a_i \sigma(a_i')$, then we calculate

$$u(\sigma)\sigma(a)u(\sigma)^{-1} = \sum_{i,j} a_i \sigma(a_i')\sigma(a)\sigma(a_j)a_j'$$

$$= P(\sigma)\left(\sum_{i,j} a_i a_i' a a_j' a_j\right) = P(\sigma)a.$$

Consequently, $P(\sigma)$ is equivalent to σ modulo the inner automorphism of A determined by $u(\sigma)$ and $P(\sigma) = \sigma$ if and only if $u(\sigma) \in Z(A)$.

We claim that $u(\sigma)P^\sigma = P$ elementwise; that is, $u(\sigma)\sigma(x) = x$ for all $x \in P$. Indeed, $u(\sigma)\sigma(x) = \sum_i a_i\sigma(a_i')\sigma(x) = \sum_i a_i\sigma(a_i'x) = \sum_i a_i a_i' x = x$. Similarly, one proves that $\sigma(x')u(\sigma)^{-1} = x'$ for all $x' \in P^{-1}$. From the fact that $u(\sigma)P^\sigma = P$ elementwise, it follows that $u(\sigma)$ is independent of the chosen decomposition of $1 \in PP^{-1}$. We now denote $u(\sigma)$ by $u(P,1,\sigma)$ to indicate the natural dependencies, and we also define: $u(P,\tau,\sigma) = \sum_i \tau(a_i)\sigma(a_i')$, which may be calculated as $u(P,\tau,\sigma) = \tau(u(P,1,\tau^{-1}\sigma))$, and this too is independent of the chosen decomposition.

I.2.4 Proposition Let $P \in I_B(A)$ and let M be any A-A-bimodule. To P we may associate an automorphism π_A in $\mathrm{Aut}_{Z(A)}(C_B(A))$ given by $\pi_A(x) = \sum_i a_i x a_i'$, where $x \in C_B(A)$ and $1 = \sum_i a_i a_i'$ is a decomposition of 1 in $PP^{-1} = B$. The map $\pi_M : C_B(M) \to C_B(M)$ defined by $\pi_M(m) = \sum_i a_i m a_i'$ for all $m \in C_B$ is a (left and right) π_A-semilinear $C_B(A)$-automorphism.

Proof: The defined maps π_A and π_M do not depend on the chosen decomposition. Indeed, let $\sum_j b_j b_j' = 1$ be another decomposition in $PP^{-1} = B$; then we have $\pi_M(m) = (\sum_i a_i m a_i')(\sum_j b_j b_j') = \sum_{i,j} a_i a_i' b_j m b_j' = \sum_j b_j m b_j'$ for any $m \in C_B(M)$. In order to prove that $\pi_M(m) \in C_B(M)$, we want $xx'\pi_M(m) = \pi_M(m)xx'$ for all $x \in P$, $x' \in P^{-1}$. Now $\pi_M(m)xx' = \sum_i a_i m a_i' x x' = \sum_i a_i a_i' x m x' = x m x' \sum_i a_i a_i' = x\sum_i x' a_i m a_i' = xx'\pi_M(m)$ indeed. If $\pi_M(m) = 0$, then $\sum a_i m a_i' = 0$; hence for all $x \in P$ we have that $\sum a_i m a_i' x = 0$. But $\sum a_i m a_i' x = \sum a_i a_i' x m = xm$, and hence $Pm = 0$ follows. Since M is an A-module, we obtain $P^{-1}Pm = 0$, and hence $m = 0$. Let us check that π_M is also surjective. Pick $y \in C_B(M)$ and let $\sum_i b_i' b_i = 1$ be a decomposition of 1 in $P^{-1}P = B$, with $b_i' \in P^{-1}$ and $b_i \in P$; put $x = \sum b_i' y b_i$. Then $\pi_M(x) = \sum_i a_i b_i' y b_i a_i' = y \sum_i a_i b_i' b_i a_i' = y$. For $r \in C_B(A)$ and $m \in C_B(M)$ we calculate that $\pi_M(rm) = \sum_i a_i r m a_i'$, whereas, on the other hand, $\pi_A(r)\pi_M(m) = (\sum_i a_i r a_i')(\sum_j a_j m a_j') = \sum_{i,j} a_i a_i' a_j r m a_j' = \sum_j a_j r m a_j' = \pi_M(rm)$. Therefore π_M is left π_A-semilinear, as claimed. The analogous property on the right is proved in a similar way. If we put $M = A$ and $\pi_M = \pi_A$, then the foregoing implies that π_A is a ring automorphism of $C_B(A)$. ■

I.2.5 Corollary Suppose we are given a group morphism $\Phi : G \to I_B(A)$, sending $g \in G$ to $P_g \in I_B(A)$, for some group G. Then for each $g \in G$ we may fix a decomposition of 1 in $P_g P_g^{-1} = B$, say $\sum a_i^g (a_i^g)' = 1$. There is a canonical action of G on $Q = C_B(A)$ given by $\psi_A(g) = \pi_A^g$, defined like π_A above, but with $P = P^g$. This action is compatible with the canonical action of G on $C_B(M)$ defined by the group morphism $\psi_M : G \to \mathrm{Aut}_{Q-Q}(C_B(M)/\psi_A)$, where $\psi_M(g) = \pi_M^g$, defined like ψ_M above, but with $P = P^g$, and where the latter group consists of the left and right ψ_A-semilinear $Q = C_B(A)$-bimodule automorphisms.

I.2.6 So we have an action of $I_B(A)$ on $C_B(A)$ given by $\psi_A(P)$ for any $P \in I_B(A)$ and also an action of $I_B(A)$ on $\mathrm{Aut}_B(A)$ given by sending σ to $P(\sigma)$ for $P \in I_B(A)$. Since $u(P,1,\sigma) \in C_B(A)$ for any $P \in I_B(A)$ and $\sigma \in \mathrm{Aut}_B(A)$, we may now easily derive the following:

I.2.7 Proposition With notation as above, for all $P \in I_B(A)$ and $\sigma, \tau \in \mathrm{Aut}_B(A)$, we have:

(a) $u(P,\tau,\sigma) = \tau(u(P,1,\tau^{-1}\sigma))$;
(b) $u(P,1,1) = 1$ for all $P \in I_B(A)$;
(c) $u(P,1,\tau\sigma) = u(P,1,\sigma)u(P^\sigma,1,\tau)$;
(d) $u(P,1,\sigma)^{-1} = u(P^\sigma,1,\sigma^{-1})$;
(e) $u(P^{\tau\sigma},1,\sigma^{-1}\tau^{-1})u(P,1,\sigma)u(P^\sigma,1,\tau) = 1$;
(f) $u(PQ,1,\sigma) = \psi_A(P)(u(Q,1,\sigma))u(P,1,\sigma)$.

Proof: (a) has been noted before; (b) is obvious.

(c) For $x \in P$ we calculate $u(P,1,\sigma)u(P^\sigma,1,\tau)\tau\sigma(x) = u(P,1,\sigma)\sigma(x) = x$ and $u(P,1,\tau\sigma)\tau\sigma(x) = x$. Since all of the appearing u's are units, the relation claimed in (c) follows.

(d) follows from (b) and (c), putting $\tau = \sigma^{-1}$; (e) follows from (c) and (d) in the obvious way.

(f) By the characterizing properties of the u's, we obtain that $u(PQ,1,\sigma)P^\sigma Q^\sigma = PQ = Pu(Q,1,\sigma)Q^\sigma$, and hence that $u(PQ,1,\sigma)P^\sigma = Pu(Q,1,\sigma) = \psi_A(P)(u(Q,1,\sigma))P$. So $u(PQ,1,\sigma)u(P,1,\sigma)^{-1}P = \psi_A(P)(u(Q,1,\sigma))P$, and thus, as the u's are all units, $u(PQ,1,\sigma) = \psi_A(P)(u(Q,1,\sigma))u(P,1,\sigma)$. ■

I.2.8 Theorem Let $P \in I_B(A)$ and assume that B is commutative. We denote $\psi_A(P)$ by σ; then for every $p' \in P^{-1}$ and $p \in P$ we have $\sigma(p'p) = pp'$. If $\sum_i a_i a_i' = 1$ is a decomposition of 1 in $PP^{-1} = B$, then also $\sum_i a_i' a_i = 1$.

Proof: Fix a decomposition $\sum_i a_i a_i' = 1$, with $a_i \in P$, $a_i' \in P^{-1}$. Since $B \subset C_B(A)$, we obtain for every $b \in B$ that $\sigma(b) = \sum_i a_i b a_i'$. Consider $p' \in P^{-1}$, $p \in P$; then we obtain the relation

$$\sigma(p'p) = \sum_i a_i p' p a_i' = \sum_i p a_i' a_i p' = p\lambda p' = pp'\lambda^\sigma. \qquad (*)$$

On the other hand, we may calculate that $pp' = \sum_i pp'a_i a_i' = \sum_i \sigma(p'a_i)pa_i'$, and also that $a_k'\sigma(p'a_i) = p'a_i a_k'$ for each $a_k' \in P^{-1}$. This then yields that $a_k'pp' = \sum_i a_k'\sigma(p'a_i)pa_i' = \sum_i p'a_i a_k' p a_i' = \sum_i p'pa_i'a_i a_k' = p'p\lambda a_k'$. For any $a_k \in P$ we obtain $a_k a_k' pp' =$

$a_k p' p \lambda a'_k$, or, summing over k,

$$pp' = \sum_i a_i p' p \lambda a'_i = \sigma(p'p)\lambda^\sigma. \qquad (**)$$

Combining $(*)$ and $(**)$ we find: $pp' = pp'(\lambda^\sigma)^2$. Since the elements of the form pp' generate B additively, the above relation leads to $(\lambda^\sigma)^2 = 1$.

Now $(pp')^2 = pp'pp' = \sigma(p'p)pp' = pp'\lambda^\sigma pp' = \lambda^\sigma (pp')^2$; consequently, $1 - \lambda^\sigma$ annihilates $(pp')^2$ for every $p \in P$, $p' \in P^{-1}$. Write $1 = \rho_1 + \cdots + \rho_r$, where $\rho_i = a_i a'_i$ and $1 = 1^{r+1} = \sum' \rho_1^{\nu_1} \ldots \rho_r^{\nu_r}$, where in each term in this sum, the exponent ν_i of some ρ_i is at least 2. Therefore, $(1 - \lambda^\sigma)\, 1 = (1 - \lambda^\sigma)(\sum' \rho_1^{\nu_1} \ldots p_r^{\nu_r}) = 0$ (since $\rho_1, \ldots, \rho_r$ commute). It follows that $\lambda^\sigma = 1$; hence $\lambda = 1$ and $(*)$ then reads as $\sigma(p'p) = pp'$. ■

We now consider left and right B-linear maps from an invertible B-bimodule M to an A-bimodule N. We say that $\varphi : M \to N$ and $\varphi' : M' \to N'$ are isomorphic if there exists a left and right B-isomorphism $f : M \to M'$ and a left and right A-isomorphism $g : N \to N'$ such that the diagram

$$\begin{array}{ccc} M & \xrightarrow{\varphi} & N \\ f\downarrow & & \downarrow g \\ M' & \xrightarrow[\varphi']{} & N' \end{array}$$

is commutative. We define $P_B(A)$ to be the set of isomorphism classes $[\varphi]$ of maps φ as above, which have the property that the induced morphism $A \otimes_B M \to N$ is an isomorphism. The product of the classes $[\varphi]$ and $[\varphi']$, where $\varphi : M \to N$ and $\varphi' : M' \to N'$, is represented by $\varphi \otimes \varphi' : M \otimes_B M' \to N \otimes_A N'$; this defines an associative multiplication on $P_B(A)$ and the inclusion map $B \subset A$ acts as the identity element. For M and N as above we put $M^* = \mathrm{Hom}_B({}_BM, {}_BB)$, $N^* = \mathrm{Hom}_A({}_AN, {}_AA)$, and we let $f^* : M^* \to N^*$

be the morphism corresponding to $f : M \to N$ in the following way: for $p^* \in M^*$ define $f^*(p^*)$ on $f(M)$ by $f^*(p^*)(f(m)) = p^*(m)$ and extend this by left A-linearity to a map of $N = A\varphi(M)$ to A. It is clear that $[\varphi^*]$ is the inverse of $[\varphi]$ and one checks that $P_B(A)$ is a group.

I.2.9 Proposition The following sequences are exact:

(a) $1 \to U(Z(A)) \cap U(Z(B)) \to U(Z(A)) \to I_B(A) \xrightarrow{\alpha} P_B(A) \xrightarrow{\beta} \mathrm{Pic}(A)$

(b) $1 \to U(Z(A)) \cap U(Z(B)) \to U(Z(B)) \xrightarrow{\gamma} \mathrm{Aut}_B(A) \xrightarrow{\delta} P_B(A) \xrightarrow{\varepsilon} \mathrm{Pic}(B)$

(c) $1 \to U(Z(B)) \to U(C_A(B)) \xrightarrow{\mu} I_B(A) \to \mathrm{Pic}(B)$

(d) $1 \to U(Z(A)) \to U(C_A(B)) \xrightarrow{\nu} \mathrm{Aut}_B(A) \to \mathrm{Pic}(A)$

Proof: (a) The maps in $1 \to U(Z(A)) \cap U(Z(B)) \to U(Z(A)) \to I_B(A)$ are the canonical ones and so far the sequence is obviously exact. For $P \in I_B(A)$ we let $\alpha(p)$ be the inclusion $P \to A$, and the latter will be isomorphic to $B \to A$ if and only if $P = Bu$ for some $u \in U(Z(A))$. To $\varphi : M \to N$ in $P_B(A)$ we associate $\beta([\varphi]) = [N]$; then $\beta([\varphi]) = 1$ leads to $N \cong A$ and $A \otimes_B M \cong A$ makes M isomorphic to a sub-B-bimodule of A, which is invertible, that is, $M \in I_B(A)$. This proves the exactness of (a).

(b) The map γ is defined by $\gamma(v)(a) = vav^{-1}$, $a \in A$, $v \in U(Z(B))$, and then it is clear that $1 \to U(Z(A)) \cap U(Z(B)) \to U(Z(B)) \to \mathrm{Aut}_B(A)$ is exact. To $\sigma \in \mathrm{Aut}_B(A)$ corresponds $\delta(\sigma) : B \to Au(\sigma) : b \to bu(\sigma)$ and for $\varphi : M \to N$ representing $[\varphi]$ in $P_B(A)$, we define $\varepsilon([\varphi]) = [M] \in \mathrm{Pic}(B)$. We leave the verification of the exactness to the reader, or refer to Y. Miyashita [45].

(c) and (d) are well known (and easy!); let us just recall that $\mu(u) = Bu$ for $u \in U(C_A(B))$, $\nu(u)(a) = uau^{-1}$ for $a \in A$. ∎

Now, suppose we are given a group morphism $\Phi : G \to I_B(A)$, where G is arbitrary, put $\Phi(\sigma) = P_\sigma$. It is clear from

$P_\sigma \subset A$ that the action $\psi_A : G \to \mathrm{Aut}(C_B(A))$ is actually an action $\psi_A : G \to \mathrm{Aut}_{Z(A)}(C_B(A))$ and by restricting to $Z(B)$, we obtain a morphism $G \to \mathrm{Aut}_{Z(A)\cap Z(B)}(Z(B))$. Moreover, G acts on $\mathrm{Pic}(B)$ by $[Q]^\sigma = [P_\sigma \otimes_B Q \otimes_B P_{\sigma^{-1}}]$. We define $\mathrm{Pic}(B)^G = \{[Q] \in \mathrm{Pic}(B); [Q]^\sigma = [Q]\}$ and $\mathrm{Pic}_{Z(B)}(B)^G = \mathrm{Pic}(B)^G \cap \mathrm{Pic}_{Z(B)}(B)$. The homomorphism $\alpha : I_B(A) \to P_B(A)$ defined in Proposition I.2.9 yields a G-action on $P_B(A)$ by taking the conjugation action induced by α. Recall from [45]:

I.2.10 Proposition The following exact sequences consist of G-morphisms:

(a) $1 \to U(Z(A)) \cap U(Z(B)) \to U(Z(B)) \to \mathrm{Aut}_B(A) \to P_B(A) \to \mathrm{Pic}(B)$;

(b) $1 \to U(Z(A)) \to U(C_A(B)) \to \mathrm{Aut}_B(A) \to \mathrm{Pic}(A)$. ■

Define $P_B(A)^G$ to consist of all $[\varphi] \in P_B(A)$, where $\varphi : M \to N$ is such that $P_\sigma \varphi(M) = \varphi(M) P_\sigma$ for all $\sigma \in G$, there is a B-module isomorphism $f_\sigma : M \to P_\sigma \otimes_B M \otimes_B P_{\sigma^{-1}}$ making the following diagram commutative:

$$\begin{array}{ccc} M & \xrightarrow{\varphi} & N \\ {\scriptstyle f_\sigma}\downarrow & & \uparrow{\scriptstyle \varphi^\sigma} \\ P_\sigma \otimes_B M \otimes_B P_{\sigma^{-1}} & & \end{array}$$

where $\varphi^\sigma(x_\sigma \otimes m \otimes y_\sigma) = x_\sigma f(m) y_\sigma$. One verifies that $P_B(A)^G$ is a subgroup of $P_B(A)$ and we define $P_Bc(A)^G = P_Bc(A) \cap P_B(A)^G$, where $P_Bc(A)$ consists of the classes of $Z(B)$-compatible bimodule maps.

I.2.11 Proposition The following sequence is exact:

$$1 \longrightarrow U(Z(A)) \cap U(Z(B)) \longrightarrow U(Z(B)) \longrightarrow \mathrm{Aut}_B(A)^G \longrightarrow P_Bc(A)^G \longrightarrow \mathrm{Pic}_{Z(B)}(B)^G,$$

where $\mathrm{Aut}_B(A)^G = \{\varphi \in \mathrm{Aut}_B(A);\ \varphi(P_\sigma) = P_\sigma \text{ for all } \sigma \in G\}$.

Proof: Follows from Proposition I.2.10 by restricting to the G-invariant and $Z(B)$-compatible parts in a straightforward way. ■

We are going to restrict attention to rings A of a particular nature, that is, to generalized crossed products over B. In this way, we obtain a generalization of the seven-term exact sequence of S. Chase, D. Harrison, and A. Rosenberg in Galois cohomology, following the path set out by Y. Miyashita in [45], although we treat generalized crossed products in the terminology of strongly graded rings, that is in the vein of C. Năstăsescu, F. Van Oystaeyen [50] or E. Dade [20]. A graded ring R of type G is said to be *strongly graded* or R is said to be a *generalized crossed product*, if for every $\sigma, \tau \in G$ we have $R_\sigma R_\tau = R_{\sigma\tau}$. If H is a subgroup of G and R is strongly graded by G, then $R^{(H)} = \oplus_{h \in H} R_h$ is strongly graded by H; if H is a normal subgroup of G, then $R_{(H)}$ is strongly graded by G/H as follows: $(R_{(H)})_{\bar{a}} = \oplus_{\sigma \in \bar{a}} R_\sigma$, $\bar{a} \in G/H$. Every graded module over the strongly graded ring R is a strongly graded module in the sense that for all $\sigma, \tau \in G$ we have $R_\sigma M_\tau = M_{\sigma\tau}$. We recall the following well-known fact [49]:

I.2.12 Preposition For any graded ring R of type G the following are equivalent:

(a) R is strongly graded.
(b) Every $M \in R$-gr is strongly graded.
(c) The functors $R \otimes \_ : R_e\text{-mod} \to R\text{-gr}$ and $(\_)_e : R\text{-gr} \to R_e$-mod define an equivalence between the Grothendieck categories R-gr and R_e-mod. ■

For $\sigma \in G$ and $M \in R$-gr, we define $M(\sigma) \in R$-gr by $(M(\sigma))_\tau = M_{\sigma\tau}$ for all $\tau \in G$.

I.2.13 Corollary If R is strongly graded by G, then for every $M \in R$-gr there is canonical graded isomorphism $M(\sigma) \cong R \otimes_{R_e} M_\sigma$. ∎

It is clear from the above results that the canonical morphism $R_\sigma \otimes R_\tau \to R_{\sigma\tau}$ is an isomorphism of R_e-bimodules, for every $\sigma, \tau \in G$. Consequently, the generalized crossed product structure of R is determined by a group morphism $\Phi : G \to \mathrm{Pic}(R_e) : \sigma \to [R_\sigma]$, together with a family F of R_e-bimodule isomorphisms $f_{\sigma,\tau} : R_\sigma \otimes R_\tau \to R_{\sigma\tau}$ satisfying the obvious associativity conditions, described by the commutativity of the following diagram (all tensors products are over R_e):

$$\begin{array}{ccc} R_\sigma \otimes R_\tau \otimes R_\gamma & \xrightarrow{R_\sigma \otimes f_{\tau,\gamma}} & R_\sigma \otimes R_{\tau\gamma} \\ {\scriptstyle f_{\sigma,\tau} \otimes R_\gamma} \downarrow & & \downarrow {\scriptstyle f_{\sigma,\tau\gamma}} \\ R_{\sigma\tau} \otimes R_\gamma & \xrightarrow[f_{\sigma\tau,\gamma}]{} & R_{\sigma\tau\gamma} \end{array}$$

Since $[R_\sigma] \in \mathrm{Pic}(R_e))$ for $\sigma \in G$, we may define an action of G on $Z(R_e)$ by letting c^σ be the element such that $R_\sigma c = c^\sigma R_\sigma$ elementwise. The group morphism $\Phi : G \to \mathrm{Pic}(R_e)$ gives rise to an action of G on $Z(R_e)$ (cf. Corollary 2.5). If we fix Φ and choose another system $\{g_{\sigma\tau}; \sigma, \tau\}$, then $g_{\sigma,\tau} = c(\sigma,\tau) f_{\sigma,\tau}$ for all $\sigma, \tau \in G$, with $c(\sigma,\tau) \in U(Z(R_e))$; this is clear, since these R_e-bimodule isomorphisms will differ by a factor that appears as the image of 1 under some R_e-bimodule isomorphism. The associativity condition for the $f_{\sigma,\tau}$ together with the action of G on $U(Z(R_e))$ represents an element of $H^2(G, U(Z(R_e)))$. Therefore, the graded isomorphism classes of the generalized crossed products corresponding to F are in bijective correspondence with the elements of $H^2(G, U(Z(R_e)))$. In this way, we obtain a group structure on the set of graded isomorphism classes of generalized crossed products corresponding to $\Phi : G \to \mathrm{Pic}(R_e)$. For more detail and a survey of some of the ring theoretical properties of generalized crossed products, we refer to [50].

Let us now construct a group of isomorphism classes $C(R/R_e)$ of generalized crossed products in a slightly more general situation. Let S be another generalized crossed product of R_e and G; then we say that S and R are R_e-*similar*, written $S \sim R$, if the R_e-bimodules S_σ and R_σ are similar.

We define $C(R/R_e)$ as the set of isomorphism classes of generalized crossed products that are similar to R. If S and T represent elements of $C(R/R_e)$, we will sometimes write them as $(S, g_{\sigma,\tau}), (T, h_{\sigma,\tau})$, if we want to specify the factor set used in the definition of the multiplication of S, resp. T. We define a product $[S] \cdot [T]$ in $C(R/R_e)$ by taking $\oplus_{\sigma \in G}(S_\sigma \otimes R_{\sigma^{-1}} \otimes T_\sigma)$ as a representative—the tensor products are over R_e.

The multiplication on the latter is defined by:

$$\begin{aligned} k_{\sigma,\tau} : S_\sigma \otimes R_{\sigma^{-1}} \otimes T_\sigma \otimes S_\tau \otimes R_{\tau^{-1}} \otimes T_\tau \\ \longrightarrow S_\sigma \otimes S_\tau \otimes R_{\tau^{-1}} \otimes R_{\sigma^{-1}} \otimes T_\sigma \otimes T_\tau \\ \longrightarrow S_{\sigma\tau} \otimes R_{(\sigma\tau)^{-1}} \otimes T_{(\sigma\tau)}, \end{aligned}$$

where the first map transposes $R_{\sigma^{-1}} \otimes T_\sigma$ and $S_\tau \otimes R_{\tau^{-1}}$ and the second map is $g_{\sigma,\tau} \otimes f_{\sigma,\tau} \otimes h_{\sigma,\tau}$. The associativity condition for $k_{\sigma,\tau}$ may be checked in a rather straightforward way (e.g., by using Theorem 1.1. of K. Hirata [30]). Obviously R will act as the identity element for $C(R/R_e)$. For the inverse of $(S, g_{\sigma,\tau})$ we may take the isomorphism class of $\oplus_{\sigma \in G}(R_\sigma \otimes S^*_\sigma \otimes R_\sigma)$ with multiplication defined by

$$\begin{aligned} h_{\sigma,\tau} : R_\sigma \otimes S^*_\sigma \otimes R_\sigma \otimes R_\tau \otimes S^*_\tau \otimes R_\tau \\ \longrightarrow R_\sigma \otimes R_\tau \otimes S^*_\tau \otimes S^*_\sigma \otimes R_\sigma \otimes R_\tau \\ \longrightarrow R_{\sigma\tau} \otimes S^*_{\sigma\tau} \otimes R_{\sigma\tau}, \end{aligned}$$

where the first map transposes $S^*_\sigma \otimes R_\sigma$ and $R_\tau \otimes S^*_\tau$ and the second is just $f_{\sigma,\tau} \otimes \pi \otimes f_{\sigma,\tau}$, where $\pi : S^*_\tau \otimes S^*_\sigma \to (S_\sigma \otimes S_\tau)^* \to S^*_{\sigma\tau}$ is the canonical isomorphism induced by $g_{\sigma,\tau}$. Therefore $C(R/R_e)$

is a group and it is abelian, as one easily checks by calculating the product both ways. By $C_{gr}(R/R_e))$ we denote the subgroup of $C(R/R_e)$ consisting of the classes of generalized crossed products $(S, g_{\sigma,\tau})$ for which $S_\sigma \cong R_\sigma$ as R_e-bimodules. By the remarks preceding the construction of $C(R/R_e)$, we see that $C_{gr}(R/R_e) = H^2(G, U(Z(R_e)))$.

Put $\mathrm{Pic}_{Z(R_e)}(R_e)^{[G]} = \{[P] \in \mathrm{Pic}_{Z(R_e)}(R_e);\ P \otimes R_\sigma \otimes {}^*P \sim R_\sigma$ for all $\sigma \in G\}$, where ${}^*P = \mathrm{Hom}_{R_e}(P, R_e)$. An abelian group $B(R/R_e)$ may then be defined by the exact sequence $\mathrm{Pic}_{Z(R_e)}(R_e)^{[G]} \to C(R/R_e) \to B(R/R_e) \to 0$. We now just mention some results of Miyashita's (cf. [45]) which lead to the generalized seven-term exact sequence, but we omit the proofs—they consist mainly of cohomological technicalities.

I.2.14 Some exact sequences The following sequences are exact:

(a) $\mathrm{Pic}_{Z(R_e)}(R)^G \to C_{gr}(R/R_e) \to B(R/R_e)$;

(b) $1 \to C_{gr}(R/R_e) \to C(R/R_e) \to Z^1(G, \mathrm{Pic}(Z(R_e))$;

(c) $C_{gr}(R/R_e) \to B(R/R_e) \to \hat{}H^1(G, \mathrm{Pic}(Z(R_e)))$, where the final term is defined by the exactness of

$$\mathrm{Pic}_{Z(R_e)}(R_e)^{[G]} \longrightarrow Z^1(G, \mathrm{Pic}(Z(R_e))) \longrightarrow \hat{}H^1(G, \mathrm{Pic}(Z(R_e))) \longrightarrow 1;$$

(d) $B(R/R_e) \to \hat{}H^1(G, \mathrm{Pic}(Z(R_e))) \to H^3(G, U(Z(R_e)))$;

(e) $1 \to U(Z(R)) \cap U(Z(R_e)) \to U(Z(R_e)) \to \mathrm{Aut}_{R_e}(R)^{(G)} \to P_{R_e}(R)^{(G)}_{Z(R_e)} \to \mathrm{Pic}_{Z(R)}(R_e)^G \to C_{gr}(R/R_e) \to B(R/R_e) \to \hat{}H^1(G, \mathrm{Pic}(Z(R_e))) \to H^3(G, U(Z(R_e)))$

(f) The exact sequence in (e) may be rewritten as:

$$1 \longrightarrow H^1(G, U(Z(R_e))) \longrightarrow (P_{R_e}(R)^{(G)}_{Z(R_e)} \longrightarrow \mathrm{Pic}_{Z(R_e)}(R_e)^G \longrightarrow H^2(G, U(R_e))) \longrightarrow B(R/R_e) \longrightarrow \hat{}H^1(G, \mathrm{Pic}(Z(R_e))) \longrightarrow H^3(G, U(Z(R_e)));$$

(g) If R_e is commutative and a Galois extension of R_e^G, the fixed ring of the action of G on R_e, then the sequence in (f) yields the well-known exact sequence:

$$\begin{aligned} 1 &\longrightarrow H^1(G,U(R_e)) \longrightarrow \mathrm{Pic}(R_e^G) \longrightarrow H^0(G,\mathrm{Pic}(R_e)) \\ &\longrightarrow H^2(G,U(R_e))) \longrightarrow Br(R/R_e) \\ &\longrightarrow H^1(G,\mathrm{Pic}(R_e)) \longrightarrow H^3(G,U(R_e)); \end{aligned}$$

(h) Recall the following version of (g) in Amitsur cohomology: let R be a commutative ring and let S be a commutative R-algebra which is an R-progenerator; then the following sequence is exact:

$$\begin{aligned} 1 &\longrightarrow H^1(S/R,U) \longrightarrow \mathrm{Pic}(R) \longrightarrow H^0(S/R,\mathrm{Pic}) \\ &\longrightarrow H^2(S/R,U) \longrightarrow Br(S/R) \\ &\longrightarrow H^1(S/R,\mathrm{Pic}) \longrightarrow H^3(S/R,U) \end{aligned}$$

where U is the units functor and Pic is the Picard functor, and where $Br(S/R) = \mathrm{Ker}(Br(R) \to Br(S))$.

We include two further results dealing with Azumaya algebras and generalized crossed products, the first of which leads to a proof of the Chase–Rosenberg sequence in (2.14.g), up to adding a few purely cohomological tricks. This result has been derived for relative Azumaya algebras by F. Van Oystaeyen in [67], Theorem 3.14, but here we present the special case of common Azumaya algebras, and for this case we should refer to T. Kanzaki [34].

I.2.15 Theorem Let A be an Azumaya algebra over $R = Z(A)$ containing a Galois extension S of R with $\mathrm{Gal}(S/R) = G$ as a maximal commutative subring; then A is a generalized crossed product of S and G, that is, $A = \oplus_{\sigma \in G} A_\sigma$, with $A_e = S$ and $A_\sigma A_\tau = A_{\sigma\tau}$ for all $\sigma, \tau \in G$.

Proof: For each $\sigma \in G$, consider the additive group

$$A_{(\sigma)} = \{u \in A;\ us = \sigma(s)u \text{ for all } s \in S\}.$$

For each $\sigma \in G$, we have that $A_{(\sigma)} \neq \emptyset$, since at least $0 \in A_{(\sigma)}$, and one easily checks that $A_{(\sigma)}$ is an S-bimodule centralizing the action of R. Indeed, for $\lambda \in S$ and $u \in A_{(\sigma)}$ we have that $(u\lambda)s = u(\lambda s) = \sigma(\lambda s)u = \sigma(\lambda)\sigma(s)u = \sigma(s)u\lambda$ and $(\lambda u)s = \lambda\sigma(s)u = \sigma(1)\lambda u$. It is obvious that for every $\sigma, \tau \in G$ we have $A_{(\sigma)}A_{(\tau)} \subset A_{(\sigma\tau)}$. If p is a prime ideal of R then after localizing at p, we reach the situation $R_p \subset S_p \subset A_p$, where A_p is an Azumaya algebra over the local ring R_p, containing a Galois extension S_p/R_p with Galois group G, as a maximal commutative subring. Since R_p and S_p have trivial Picard groups, the classical crossed product result (cf. DeMeyer and E. Ingraham [23]) yields that $A_p = \oplus_{\sigma \in G} S_p \cdot u_\sigma$, where each u_σ is a unit of A_p, inducing σ in S_p by conjugation. Let $\kappa_p A$ be the torsion part of A at p. For some $t \in R - p$, we have that $tu_\sigma \in A/\kappa_p A$. Pick $y \in A$ such that $j_p(y) = tu_\sigma$, where $j_p : A \longrightarrow A/\kappa_p A$ is the canonical morphism. Then $ys - \sigma(s)y \in \kappa_p A$ for every $s \in S$, and since S is finitely generated, we can annihilate $ys - \sigma(s)y$ by some $\lambda \in R - p$ which does not depend on s, that is, $\lambda y \in A_{(\sigma)}$ or $\lambda t u_\sigma \in S_p A_{(\sigma)} = (A_{(\sigma)})_p$. From $\lambda t \in R - p$, it then follows that $u_\sigma \in S_p A_{(\sigma)}$ and consequently $A_p = \oplus_{\sigma \in G}(A_{(\sigma)})_p$, with $(A_{(\sigma)})_p = S_p u_\sigma$. For $\sigma, \tau \in G$, $A_{(\sigma)} \cap A_{(\tau)}$ localizes to zero at each prime ideal p of R, because $(A_{(\sigma)})_p \cap (A_{(\tau)})_p = S_p u_\sigma \cap S_p u_\tau = 0$; consequently $A_{(\sigma)} \cap A_{(\tau)} = 0$. On the other hand, $B = \oplus_{\sigma \in G} A_{(\sigma)}$ localizes to A_p at each prime ideal p of R, hence $A/B = 0$ or $A = B$. Finally $A_{(\sigma)}A_{(\tau)}$ and $A_{(\sigma\tau)}$ both localize to $(A_{(\sigma\tau)})_p$ at each time prime ideal p of R, hence $A_{(\sigma\tau)} = A_{(\sigma)}A_{(\tau)}$ for all $\sigma, \tau \in G$. So, putting $A_\sigma = A_{(\sigma)}$ and $A_e = S$ defines a strong gradation on A. ■

I.2.16 Corollary Viewing the A_σ as representatives of $[A_\sigma] \in \text{Pic}(S)$, the reader will easily convince himself that the above theorem states exactly that the following sequence is exact:

$$H^0(G,\mathrm{Pic}(S)) \longrightarrow H^2(S/R,U(S)) \longrightarrow Br(S/R) \longrightarrow H^1(G,\mathrm{Pic}(S)).$$

The remaining terms of the seven-term exact sequence may be added; this requires only some elementary cohomological techniques.

The connection between Galois actions and strong gradations may be dualized in a perhaps surprising way, as the following shows.

I.2.17 Proposition If A is an algebra with center C such that A contains a G-strongly graded ring R over C, that is, with $R_e = C$, then there is a natural action of G on A such that $A^G = C_A(R)$.

Proof: Since A is an R-bimodule such that $Z_A(R_e) = A$, we may define the action of G on the R-bimodule A as before, say $r_\sigma a = \psi_A(\sigma)(a)r_\sigma$ for every $r_\sigma \in R_{\sigma'}$ $\sigma \in G$, and $a \in A$, or by using a decomposition $1 = \sum_i u_{i,\sigma}v_{i,\sigma^{-1}}$ with $u_{i,\sigma} \in R_\sigma$ and $v_{i,\sigma^{-1}} \in R_{\sigma^{-1}}$ as $\psi_A(\sigma)(a) = \sum_i u_{i,\sigma}av_{i,\sigma}av_{i,\sigma^{-1}}$. In this case, one easily verifies that $\psi_A(\sigma)$ is a ring homomorphism, actually even an R_e-algebra morphism. If $\psi_A(\sigma)(a) = a$ for all $\sigma \in G$ and some $a \in A$, then a commutes with every element of R, hence $A^G = C_A(R)$ follows. ■

The above result remains valid if R is assumed only to be a *Clifford system* for G, that is, $R = \sum_{\sigma \in G} R_\sigma$, with $R_\sigma R_\tau = R_{\sigma\tau}$ for all $\sigma, \tau \in G$; the proof is exactly the same. Let us point out the following results, which are modifications of the foregoing.

I.2.18 Proposition Let A be a k-central simple algebra containing a simple algebra B; then the following statements are equivalent:

(1) $B = A^G$ for some group G acting on A by automorphisms;
(2) $C_A(B)$ is a Clifford system for G over k.

Proof: Starting from (2.18.2), we may define the G-action on A as in the foregoing proposition and then $A^G = C_A(C_A(B))) = B$. Starting from (2.18.1), we identify G with the subgroup of $\mathrm{Aut}_k(A)$ it determines (this is no loss of generality, since we have phrased the result in terms of Clifford systems). For each $\sigma \in G$, there is an $u_\sigma \in A$ such that $\sigma(a) = u_\sigma^{-1} a u_\sigma$ for all $a \in A$. Since $u_\sigma u_\tau u_{\sigma\tau}^{-1}$ determines the trivial inner automorphism of A, it follows that $u_\sigma u_\tau = k(\sigma,\tau) u_{\sigma\tau}$ with $k(\sigma,\tau) \in k^*$. The k-vector space D generated by $\{u_\tau; \tau \in G\}$ is a k-algebra. Since B is fixed under the conjugations determined by the u_σ, it follows that $k \subset D \subset C_A(B)$, but then also $B = C_A(D)$, because $C_A(D) = A^G$. By the double commutator theorem, it follows that $D = C_A(B)$. By construction D is a G-Clifford system over k, with $D_\sigma = k u_\sigma$, $\sigma \in G$. ■

I.2.19 Corollary If A is an Azumaya algebra with center C, such that A contains a maximal commutative subring R, which is a Clifford system for G over C, that is, with $R_e = C$, then there is a natural action of G on A such that $A^G = R$. ■

This corollary yields a *dual crossed product theorem* for strongly graded splitting rings, because in case R is strongly graded it actually states that $A = R^\sharp(CG)^*$, where $(CG)^*$ is the C-dual of CG.

Chapter II

Some Results on Commutative Graded Rings

II.1 ARITHMETICALLY GRADED RINGS

Let G be an abelian group and consider a commutative ring R graded by G. The so-called *arithmetically graded* rings focused on in this section will be Krull domains graded by G, but we expound only those properties that are relevant for applications in the study of graded Azumaya algebras over the rings we consider. It will be proved in a subsequent section that for any Azumaya algebra which is G-graded over R, there exists a noetherian subring R' of R and an Azumaya algebra A' over R' such that $A = A' \otimes_{R'} R$ and A' and R' are G-graded. Checking the proof of this result, it shows that R' is actually finitely generated by homogeneous elements over the prime ring of R, hence R' is graded by some finitely generated subgroup of G. Since A' is a finite R'-module, it also follows that A' is graded by some finitely generated subgroup H of G. Since $A' \subset A^{(H)}$ and

$R' \subset R^{(H)}$, it follows that $A^{(H)} = A' \otimes_{R'} R^{(H)}$ and thus that $A^{(H)}$ is an Azumaya algebra over $R^{(H)}$, graded by H. This establishes that the study of G-graded Azumaya algebras over R may be reduced to the study of H-graded Azumaya algebras over the ring $R^{(H)}$, for some finitely generated abelian group H. Since $H = Z \times H_{\text{tors}}$, where H_{tors} is finite, it follows that only the cases where $G = \mathbf{Z}^m$ and G is finite have to be dealt with. Here we specialize to the torsion free case and state some results for torsion free abelian groups, if the restriction to the case $G \cong \mathbf{Z}^m$ is not essential.

Consider a torsion free abelian group G and let $\Gamma \subset G$ be some submonoid such that the group $\langle \Gamma \rangle$ generated by Γ is exactly G. Since G is ordered, it will sometimes be useful to let Γ be the set of nonnegative elements of G in the ordering considered; for most of the applications, however, we will have $\Gamma = G$.

Let R be a commutative domain graded by Γ. The set $S = \{x \neq 0 : x \in h(R)\}$ is multiplicatively closed and $S^{-1}R$ is G-graded such that $(S^{-1}R)_0$ is a field and every nonzero $y \in h(S^{-1}R)$ is an invertible element of $S^{-1}R$. We write $Q^g = S^{-1}R$ and we say that Q^g is a graded field. There is an obvious exact sequence:

$$1 \longrightarrow ((Q^g)_0)^* \longrightarrow (Q^g)^* \longrightarrow G \longrightarrow 1$$

(up to replacing G by the subgroup of degrees really appearing in Q^g, but it is not restrictive at all if we assume that G itself is that group). It is obvious that $(Q^g)_\sigma = (Q^g)_0 u_\sigma$ for any $u_\sigma \neq 0$ in $(Q^g)_\sigma$ and, once we fix such representatives, the elements u_σ satisfy $u_\sigma u_\tau = t(\sigma,\tau) u_{\sigma\tau}$ for all $\sigma, \tau \in G$. It is easily verified that $t : G \times G \to ((Q^g)_0)^*$ is a 2-cocyle and hence $Q^g = (Q^g)_0 G^t$ is the twisted group ring with respect to the cocyle t. If G is a free group (as we may assume in the study of Azumaya algebras, in view of the introductory remarks), then $Q^g = (Q^g)_0 G$ and this is true for general torsion free abelian groups, whenever $(Q^g)_0$ is root-closed (i.e., if one extracts an nth root of an element in $(Q^g)_0$, then this root is in $(Q^g)_0$).

II.1.1 Proposition The following assertions are valid:
(1.1.1) Q^g is completely integrally closed.
(1.1.2) Finite intersections of principal ideals of Q^g are again principal.
(1.1.3) Q^g is a Krull domain if and only if it satisfies the ascending chain condition with respect to principal ideals, if and only if Q^g is factorial.
(1.1.4) If G satisfies the ascending chain condition on cyclic subgroups, then Q^g is a Krull domain (the converse is false in general!).
(1.1.5) If Γ satisfies the ascending chain condition on cyclic submonoids, then Q^g is a factorial Krull domain.
(1.1.6) If Q^g is factorial and H is any subgroup of G, then $(Q^g)^{(H)}$ is factorial again.
(1.1.7) The Γ-graded domain R is integrally closed if and only if R_0 is integrally closed in $(Q^g)_0$.

Proof: We leave this as an exercise; combining properties established in R. Matsuda [43], D. D. Anderson, D. F. Anderson [1], or F. Van Oystaeyen [42], the result follows. ∎

II.1.2 Remark Let Γ be the set of nonnegative elements in an ordering of G and write $(Q^g)_+ = \oplus_{\nu\in\Gamma}(Q^g)_\nu$. Clearly Q^g is the graded ring of fractions of $(Q^g)_+$ and from the above proposition, it is not hard to deduce that $(Q^g)_+$ is integrally closed.

If we assume that R is an integrally closed domain, then $\mathrm{Cl}(R) = \mathrm{Cl}^g(R)$ holds if and only if each divisorial ideal I of R becomes principal in Q^g. Consequently, $\mathrm{Cl}^g(R) = \mathrm{Cl}(R)$ holds, for example, if $G \cong \mathbf{Z}^m$, $G = (\mathbf{Z}^m)_+$, or in case R

is a Krull domain. For results on class groups of Krull domains, we refer to R. Fossum [25], or F. Van Oystaeyen and A. Verschoren [69]. We include a lemma, due to C. Weibel [73], which is useful in certain situations for the study of Brauer groups:

II.1.3 Lemma Let R be a commutative ring graded by an abelian cancelative monoid Γ with only the trivial unit. If F is a functor from *Rings* to *Groups*, such that $R \to R[\Gamma]$ induces an isomorphism $F(R) \cong F(R[\Gamma])$, then it also induces an isomorphism $F(R_0) \cong F(R)$.

Proof: Consider the following homomorphisms:

$$R \xrightarrow{i} R[\Gamma] \xrightarrow{\pi_i} R \qquad (i = 1, 2)$$

where $\pi_1(\sum r_\sigma \sigma) = \sum r_\sigma$ and $\pi_2(\sum r_\sigma \sigma) = r_0$; then $F(i)$ is an isomorphism by assumption and since $\pi_1 i = \pi_2 i = 1_R$, it follows that both $F(\pi_1)$ and $F(\pi_2)$ are isomorphisms. Define $f : R \to R[\Gamma]$ by $f(\sum r_\sigma) = \sum r_\sigma \sigma$; then $\pi_1 f = 1_R$, hence $F(f)$ is an isomorphism and $\pi_2 f = \varepsilon$ is the augmentation $R \to R_0$, and hence $F(\varepsilon)$ is an isomorphism. From the fact that the composition $R_0 \to R \to R_0$ (the last map is the augmentation!) yields the identity on R_0, it follows that $F(R_0) \to F(R)$ is an isomorphism. ■

II.1.4 Corollary If R is a Krull domain graded by a monoid Γ as above, then $\mathrm{Pic}(R_0) \cong \mathrm{Pic}(R)$. If A is a regular ring which is affine over a field, then projective $A[T]$-modules are extended from projective A-modules. If A is positively graded, then it follows that projective A-modules extend from projective A_0-modules (well known, because such rings have to be symmetric algebras of a finitely generated projective A_0-module). ■

Recall that if R is strongly graded by G, then the map $\sigma \to [R_\sigma]$ defines a group homomorphism $\varphi : G \to \mathrm{Pic}(R_0)$. It is rather straightforward to establish:

II.1.5 Lemma Let the commutative Krull domain R be strongly graded by the torsion free abelian group G; then we obtain the following exact sequences of abelian groups:

$$1 \longrightarrow \mathrm{Im}\,\varphi \longrightarrow \mathrm{Pic}(R_0) \longrightarrow \mathrm{Pic}(R) \longrightarrow 1;$$

$$1 \longrightarrow \mathrm{Im}\,\varphi \longrightarrow \mathrm{Cl}(R_0) \longrightarrow \mathrm{Cl}(R) \longrightarrow 1. \qquad \blacksquare$$

The foregoing lemma may be formulated for general graded domains, if one uses a suitable definition for the class "group" (i.e., monoid) in that case, but we will be interested mostly in this kind of result when R_0 is (completely) integrally closed. In this context, it is very natural to modify the definition of a strongly graded ring in order to obtain a notion more intrinsically linked to the class group. It is restrictive to have that $\mathrm{Ker}(\mathrm{Cl}(R_0) \to \mathrm{Cl}(R)) = \mathrm{Ker}(\mathrm{Pic}(R_0) \to \mathrm{Pic}(R))$, particularly if we aim to study Krull domains, which are not necessarily locally factorial (if R is locally factorial, then $\mathrm{Pic}(R) = \mathrm{Cl}(R)$ and everything is very easy!).

We say that a domain graded by G is *divisorially graded* if $R = \oplus_{\sigma \in G} R_\sigma$ with R_σ being a divisorial R_0-lattice such that $(R_\sigma R_\tau)^{**} = R_{\sigma\tau}$ for all $\sigma, \tau \in G$. It is obvious that a divisorial gradation corresponds to a monoid morphism $\Phi : G \to \mathrm{Cl}(R_0)$, where $\mathrm{Im}(\Phi)$ is a group, of course, together with a factor set obtained from the isomorphisms $R_\sigma \perp R_\tau \cong R_{\sigma\tau}$ (where $\perp$ is the modified tensor product $(\_ \otimes \_)^{**}$—all tensorproducts over R_0). A strongly graded ring is a particular example of a divisorially graded ring.

II.1.6 Theorem Let the commutative domain R be divisorially graded by a torsion free abelian group G.
(1.6.1) If R_0 is integrally closed in its field of fractions Q_0, then R is integrally closed in its field of fractions Q.
(1.6.2) If G satisfies the ascending chain condition on cyclic subgroups and R_0 is a Krull domain, then R is a Krull domain too.

Proof: See [67]. ■

Because for all $\sigma \in G$ the R_σ are divisorial R_0-lattices, instead of $(\_)^{**}$ one may use the functor $Q_{1,0}$, the localization functor corresponding to the idempotent functor associated to $X^1(R_0)$, that is, $\kappa_1^0 = \inf\{\kappa_p : p \in X^1(R_0)\}$. Then the definition of divisorial gradation on R coincides with the notion of κ_1-graded ring, introduced in [15]. As a particular case of Lemma 1.2.1 and Proposition 1.2.3 of loc. cit. we obtain:

II.1.7 Lemma Let R be divisorially graded by the torsion free abelian group G over the Krull domain R_0; then for any graded left ideal L of R, we have $Q_{1,0}(L) = Q_{1,0}(RL_0)$. If M is a κ_1^0-closed R-module, then $Q_{1,0}(R \otimes M_0) \cong_g M$. ■

We are now ready to prove:

II.1.8 Theorem Let R be a Krull domain, divisorially graded by a torsion free abelian group G; then the following assertions are valid:
(1.8.1) R_0 is a Krull domain.
(1.8.2) The extension of Krull domains $R_0 \to R$ satisfies the PDE condition.
(1.8.3) The map $\pi : \mathrm{Cl}(R_0) \to \mathrm{Cl}(R)$ defined by $\pi(I) = [Q_{1,0}(RI)]$ is a group morphism; consequently, we have for every ideal L of

R that $Q_{1,0}(L) = Q_1(L)$, the localization at $\kappa_1 = \inf\{\kappa_p : P \in X^1(R)$.

(1.8.4) There is an exact sequence of abelian groups

$$I \longrightarrow \mathrm{Im}(\Phi) \longrightarrow \mathrm{Cl}(R_0) \longrightarrow \mathrm{Cl}(R) \longrightarrow 1,$$

where $\Phi : G \to \mathrm{Cl}(R_0)$ is a group morphism associated to the divisorial gradation on R.

Proof: (1) is obvious, because $R_0 = R \cap (Q^g)_0$.

To prove (2), take $P \in X^1(R)$; then either $P_g = 0$ or $P = P_g$ (because G is ordered, P_g is prime!—recall that P_g is the largest graded ideal contained in P). If $P_g = 0$, then $P \cap R_0 = 0$ or $\mathrm{ht}(P \cap R_0) < 1$. If $P_g = P$, let $p = P \cap R_0$. By the Lemma, we have $P = Q_{1,0}(Rp)$. If $p \notin X^1(R_0)$, then we can find some $q \in X^1(R_0)$ with $q \subset p$. We claim that $J = Q_{1,0}(Rq)$ is a graded prime ideal of R. That J is graded is easily deduced from the fact that Rq is graded and $Q_{1,0}$ is a localization in R_0-mod (actually, if $x = x_{\sigma(1)} + \cdots + x_{\sigma(n)} \in J$, then $Ix \subset Rq$ for some $I \in L(\kappa_1^0)$, the Gabriel filter associated to κ_1^0, hence $Ix_{\sigma(1)} \subset Rq, \ldots, Ix_{\sigma(n)} \subset Rq$ or $x_{\sigma(i)} \in Q_{1,0}(Rq)$ for $i = 1, \ldots, n$). Suppose that l_1 and l_2 are homogeneous elements of R, such that $l_1 l_2 \subset Q_{1,0}(Rq)$; then $I' l_1 l_2 \subset Rq$ for some $I' \in L(\kappa_1^0)$. Consequently, $I'(R_{\sigma(1)} l_1)(R_{\sigma(2)} l_2) \subset q$ if $\deg(l_1) = \sigma(1)^{-1}$ and $\deg(l_2) = \sigma(2)^{-1}$. Since $q \in X^1(R_0)$, $I' \not\subset q$, hence either $R_{\sigma(1)} l_1 \subset q$ or $R_{\sigma(2)} l_2 \subset q$, say $R_{\sigma(1)} l_1 \subset q$, then $R_{\sigma(1)^{-1}} R_{\sigma(1)} l_1 \subset Rq$, and as $R_{\sigma(1)^{-1}} R_{\sigma(1)} \in L(\kappa_1^0)$, it follows from the latter that $l_1 \in Q_{1,0}(Rq)$. The minimality assumption on P yields $Q_{1,0}(Rq) = P = Q_{1,0}(Rp)$. If $p \neq q$, pick $a \in P - q$. The foregoing equality yields that $I'' a \subset Rq$ for some $I'' \in L(\kappa_1^0)$, hence $I'' a \subset q$ with $I'' \not\subset q$, or $a \in q$, a contradiction. Consequently, $p \in X^1(R_0)$ or $\mathrm{ht}(P \cap R_0) < 1$ holds for all $P \in X^1(R)$, which proves (2).

In order to verify (3), first note that the divisor group of a Krull domain is generated by the prime ideals of height one and since $\mathrm{Cl}(R) = \mathrm{Cl}^g(R)$, it is obvious that (3) follows from (2).

Indeed, from the second statement, we have $Q_{1,0}(L) = Q_{1,0}(RL_0)$ and also $Q_{1,0}(L) = Q_{1,0}(RQ_{1,0}(L_0))$, hence we may decompose $Q_{1,0}(L_0) = L_0 = p_1^{r(1)} * \cdots * p_m^{r(m)}$ in $\mathrm{Div}(R_0)$ and obtain $Q_{1,0}(L) = P_1^{r(1)} * \cdots * P_m^{r(m)}$ in $\mathrm{Div}(R)$ by extending, where $P_1^{r(1)} * \cdots * P_m^{r(m)}$ obviously coincides with $Q_1(L)$, since it is the smallest R-divisorial module containing L.

Finally, to prove (4), if I represents an element $[I]$ in $\mathrm{Cl}(R)$, then $I = Q_{1,0}(Rl_0)$; consequently π is surjective. On the other hand, if $[J] \in \mathrm{Im}(\Phi)$, that is, $[J] = [R_\sigma]$ for some $\sigma \in G$, then $Q_{1,0}(RJ) = Q_{1,0}(RR_{\sigma^{-1}}J) = L$ is such that $[L] = 1$, because $[Q_{1,0}(R_{\sigma^{-1}}J)] = 1$ in $\mathrm{Cl}(R_0)$. Moreover, if $[I] \in \mathrm{Cl}(R_0)$ is mapped to 1 by π, then $Q_{1,0}(RI) = Ra$ for some homogeneous $a \in R$ of degree $\tau \in G$, say. Taking parts of degree 0 yields that $I = R_{\tau^{-1}}a$, hence $[I] = [R_{\tau^{-1}}] \in \mathrm{Im}(\Phi)$. ■

At this point, we restrict to the case $G = \mathbf{Z}$ and study a very interesting class of graded Krull domains (of dimension 2), consisting of the gr-Dedekind domains. The latter rings are important as examples, but also as constructive tools. In the theory of Brauer groups of graded rings, these gr-Dedekind rings are particularly well behaved, although their structure and that of their Brauer groups is by no means trivial.

We consider a $\mathbf{Z}$-graded gr-field $K^g \cong k[T,T^{-1}]$. A graded subring O_v of K^g is a *gr-valuation ring* of K^g if for every $x \in h(K^g)$ either $x \in O_v$ or $x^{-1} \in O_v$.

II.1.9 Theorem Let O_v be a gr-valuation ring of K^g; then the following assertions are valid:

(1.9.1) The graded ideals of O_v are linearly ordered by inclusion.

(1.9.2) For $x_1, \ldots, x_n \in h(K^g) - O_v$, we have $(x_i)^{-1}x_j \in O_v$ for some $i \in \{1, \ldots, n\}$ and all $j \in \{1, \ldots, n\}$.

(1.9.3) O_v is a gr-local ring; let M_v be the unique gr-maximal ideal of O_v. We may associate to O_v a valuation function $v : (K^g)^* \to \Gamma$

for some ordered group Γ. This function v extends to a valuation $v : K^* \to \Gamma$, where K is the field of fractions of K^g. If $x \in K^g$, then $x \in O_v$ if and only if $v(x) \geq 0$ and $x \in M_v$ if and only if $v(x) > 0$.

Proof: See [49], BI.3.2, BI.3.3 and BI.3.13. ■

If v is a gr-valuation with $\Gamma = \mathbf{Z}$, then we say that v is a *discrete* gr-valuation or a *principal* gr-valuation. Since a $\mathbf{Z}$-graded ring R is noetherian if and only if it is gr-noetherian (cf. [48], II.3.1), we may easily prove:

II.1.10 Proposition The gr-valuation ring O_v is noetherian (gr-noetherian) if and only if v is discrete. ■

A discrete gr-valuation ring will be a noetherian integrally closed domain of graded Krull dimension exactly one and therefore Krull $\dim(O_v) \leq 2$. If $P \in X^1(O_v)$, then either $P = M_v$ or $P_g = 0$. If $P \neq M_v$, then $(O_v)_P$ is a localization of K^g at PK^g and therefore it is a discrete valuation ring of K (the associated valuation on K does not come from a graded valuation on K^g).

The only discrete gr-valuation rings of $k[T,T^{-1}]$ of Krull dimension one are $k[T]$ and $k[T^{-1}]$; this may be proved very easily, but is also follows from the following structure theorem:

II.1.11 Theorem (M. Van den Bergh) A discrete gr-valuation ring R of $k[T,T^{-1}]$ (with $\deg(T) = 1$) is necessarily of one of the following types:

(1.11.1) $R = k[T]$;

(1.11.2) $R = k[T^{-1}]$;

(1.11.3) $E = \sum_{n \in \mathbf{Z}} M_0^{-na} T^n$, where $a \in \mathbf{Q}$ and M_0 is the maximal ideal of a discrete valuation ring R_0 of k.

Proof: Let $v : h(K^g) \to \mathbf{Z}$ be the graded valuation corresponding to R. First, $v(k) = 0$ entails that $k \subset R$ and either $k[T] \subset R$ or $k[T^{-1}] \subset R$. Both rings mentioned turn out to be discrete gr-valuation rings; hence they are maximal graded subrings of $k[T,T^{-1}]$ and we arrive at the possibilities (1) and (2). Next, consider the case $v(k) = n\mathbf{Z}$ for some positive integer n; then R_0 is a discrete valuation ring of k with valuation $v|_k$. We may normalize v so that $v(k) = \mathbf{Z}$ (in the sequel we will assume all valuations to be normalized this way), and then we put $v(T) = a$. By the definition of v, we have $h(R) = \{x \in h((K^g)^*); v(x) \geq 0\} = \{yT^i : y \in k^*, v(yT^i) \geq 0, i \in \mathbf{Z}\} \cup \{0\}$ and this is just $\sum_{i \in \mathbf{Z}} M_0^{-ia} T^i$, where $M_0^a = \{y \in k^* : v(y) \geq a\}$. ■

Up to replacing T by $\pi^m T$, where π is a uniformizing parameter of R_0, we may assume that $0 \leq a < 1$. The *type* of R, denoted by $t(R)$, is the number $a \bmod 1$. Obviously, R_0 and $t(R)$ completely determine the discrete gr-valuation ring R.

II.1.12 Proposition Let R be a discrete gr-valuation ring of $k[T,T^{-1}]$ of type $t(R) = p/e$, where $0 < p < e$ and $(p,e) = 1$ (if $t(R) = 0$, put $p = 0$ and $e = 1$). Let m be the maximal ideal of R_0 and M the gr-maximal ideal of R. Then
(1.12.1) $M^e = Rm$ (we call e the ramification index of R_0 in R).
(1.12.2) The units of R are homogeneous of degree he, with $h \in \mathbf{Z}$.
(1.12.3) Uniformizing elements of R have degrees p' with $pp' = 1 \bmod e$.
(1.12.4) $R/M \cong_g R_0/m[T^e, T^{-e}]$, with $\deg T = 1$.
(1.12.5) If $x \in h(K^g)$, then $v(x) = a \deg(x) \bmod 1$.

Proof: Elementary, if we use the structure result (1.11); details are given in [57], 3.4. ■

II.1.13 Remark Property (II.1.5) of the above result implies that there are elements with negative valuation of nonzero degree.

A **Z**-graded commutative domain is called a *gr-principal ideal domain* if every graded ideal of it is principal. A **Z**-graded domain is said to be a *gr-Dedekind* ring if every graded ideal of it is projective. The following results may be proved just like their ungraded equivalents:

II.1.14 Theorem For any **Z**-graded domain R, the following assertions are equivalent:
(1.14.1) R is a gr-Dedekind ring.
(1.14.2) R is a Krull domain and nonzero graded prime ideals of R are gr-maximal.
(1.14.3) R is noetherian and integrally closed and graded prime ideals of R are gr-maximal.
(1.14.4) Every graded ideal of R is invertible.
(1.14.5) Every graded ideal of R is in a unique way a product of graded prime ideals.
(1.14.6) R is noetherian, homogeneously integrally closed in K^g, and nonzero graded prime ideals are gr-maximal.
(1.14.7) The graded fractional ideals of R form a multiplicative group.
(1.14.8) R is noetherian and each localization R_m at a gr-maximal ideal m of R is a principal ideal domain.
(1.14.9) R is noetherian and each graded localization R_p^g at a graded prime ideal p of R is a gr-principal ideal domain.
(1.14.10) R is noetherian and each graded localization R_m^g at a gr-maximal ideal m of R is a gr-principal ideal ring.
(1.14.11) All graded R-modules which are gr-divisible are injective in R-gr. ■

II.1.15 Proposition Let O_v be a gr-valuation ring K^g; then the following assertions are equivalent:
(1.15.1) O_v is a discrete gr-valuation ring;
(1.15.2) O_v is a gr-Dedekind domain;
(1.15.3) O_v is a gr-principal ideal domain;
(1.15.4) M_v is generated by a homogeneous element of O_v;
(1.15.5) O_v is factorial;
(1.15.6) O_v satisfies the ascending chain condition on principal ideals;
(1.15.7) O_v is a Krull domain and $\mathrm{ht}(M_v) = 1$. ∎

II.1.16 Lemma Let R be a gr-Dedekind domain which is not of the form $R_0[T]$. Then:
(1.16.1) R_0 is a Dedekind domain.
(1.16.2) There is a positive integer e such that $R = \oplus_{n\in\mathbf{Z}} R_{en}$, with $R_e \neq 0$.
(1.16.3) Every fractional graded ideal of R may be generated by two homogeneous elements, one of which may be chosen arbitrarily in the ideal.

Proof: See [49], BII.2.3–4. ∎

Since a gr-Dedekind domain is a graded Krull domain, we know that $\mathrm{Pic}(R) = \mathrm{Pic}^g(R)$, $\mathrm{Cl}(R) = \mathrm{Cl}^g(R)$, but now $\mathrm{Pic}^g(R) = \mathrm{Cl}^g(R)$, because all graded fractional ideals of R are projective. Hence $\mathrm{Pic}(R) = \mathrm{Cl}(R)$, although R may (and usually will) have Krull dimension 2. If R is a strongly graded gr-Dedekind domain, let $\Phi : \mathbf{Z} \to \mathrm{Pic}(R_0)$ be the map given by sending 1 to I, that is, R is the generalized Rees ring $D^{\sim}(I) = \sum_{n\in\mathbf{Z}} I^n X^n$ and $\mathrm{Cl}(R) = \mathrm{Cl}(R_0)/\langle [I] \rangle$. One verifies that $D^{\sim}(J) \cong D^{\sim}(I)$ as graded rings if and only if $[I] = [J]$ in $\mathrm{Cl}(R_0)$. Recall that for any $\mathbf{Z}$-graded ring A, we write $A^{(e)}$ for $A^{(\mathbf{Z}e)} = \oplus_{m\in\mathbf{Z}} A_{me}$, equipped with the gradation $(A^{(e)})_m = A_{em}$.

II.1.17 Lemma (1.17.1) If R is a gr-Dedekind domain, then for all positive integers e, the ring $R^{(e)}$ is a gr-Dedekind domain. (1.17.2) If P is a graded prime ideal of a gr-Dedekind domain R, then R^g_P, the graded ring of fractions at P, is obtained by localizing at $R_0 - P_0$.

Proof: For (2) we have seen that R^g_P is a discrete gr-valuation ring. Now R/P is a gr-field. If $z \in (R/P)_n$, then there is an inverse $z^{-1} \in (R/P)_{-n}$ and if y represents z (resp., y' represents z^{-1}), then it suffices to invert $yy' \in R_0$ in order to invert $y \notin P$. It follows that $R^g_P = R_{0,P_0} \otimes R$, and (1) follows from (1.11) and (1.14.9). ■

II.1.18 Theorem Let R be a gr-Dedekind domain; then there exists a positive integer e and a fractional ideal I of R_0 such that $R^{(e)} = R_0^{\sim}(I)$.

Proof: Consider the graded prime ideals dividing RR_1 and let e be the least common multiple of their ramification indices over R_0; then one calculates that $R^{(e)}$ is strongly graded, because no graded ideal of $R^{(e)}$ contains $(R^{(e)})_1 = R_e$. ■

Now consider an integrally closed domain R_0 and an invertible ideal I of R_0. For any integer p, we define $I^{1/p}$ to be the sum of all ideals J of R_0 satisfying $J^p \subset I$. One may check that $(I^{1/p})^p \subset I$. For $a = p/q$, we define I^a as $(I^p)^{1/q}$; this makes sense, because $(I^p)^{1/q} = (I^{np})^{1/nq}$ for all positive integers n. Over a Dedekind domain R_0, the foregoing definitions reduce to the classical ones in terms of invertible ideals. It is equally easy to check that the following properties hold for an invertible ideal in an integrally closed domain R_0:

(1) For any $a, b \in \mathbf{Q}$, we have $I^a I^b \subset I^{a+b}$.
(2) For any $a \in \mathbf{Q}$, we have $I^a J^a \subset (IJ)^a$.
(3) For any $a, b, c \in \mathbf{Q}$ such that $a < b < c$, we have $I^a \cap I^c \subset I^b$.

We are now ready to mention M. Van den Bergh's structure result for gr-Dedekind domains (cf. [57]):

II.1.19 Theorem Let R be a $\mathbf{Z}$-graded integrally closed ring such that $R^{(e)}$ is a generalized Rees ring for some positive integer e; then $R = \sum_{n \in \mathbf{Z}} I^{n/e} X^n$, where I is invertible. Conversely, if R_0 is integrally closed, then every graded ring of this type is also integrally closed.

Proof: See loc. cit. Note that the second statement also follows from an elementary elaboration of Theorem (II.1.6). ∎

If R is a gr-Dedekind domain, then there exists a bijective correspondence between maximal ideals of R_0 and gr-maximal ideals of R. If $M \in \mathrm{Max}^g(R)$, the set of gr-maximal ideals of R, and $m = M \cap R_0$, then $R^g_M = (R_0)_m \otimes R$. Denote the corresponding gr-valuation (resp. ramification index) by v_M (resp. e_M). We say that R satisfies the *graded approximation property* (GAP) if for any finite subset S of $\mathrm{Max}^g(R)$ and any set of integers $\{n_M : M \in S\}$, there exists an $x \in h(K^g)$ such that $e_M v_M(x) = n_M$ for all $M \in S$ and $v_M(x) \geq 0$ for $M \notin S$.

II.1.20 Theorem A gr-Dedekind ring R satisfies GAP if and only if for every $P \neq Q$ in $\mathrm{Max}^g(R)$, we have that $(e_P, e_Q) = 1$. ∎

II.1.21 Proposition Let R be a $\mathbf{Z}$-graded gr-Dedekind ring.
(1.21.1) If $M \in R$-gr is finitely generated, then $M \cong N \oplus T$, where $N \in R$-gr is torsion free and $T \in R$-gr is torsion.
(1.21.2) For T as in (1) we have $T \cong \oplus R/(P_i)^{n(i)}$, with $P_i \in \mathrm{Max}^g(R)$.
(1.21.3) If M is a graded R-lattice, then $M = I_1 \oplus \cdots \oplus I_n$, where $I_1, \ldots, I_n$ are graded fractional ideals. ∎

For any **Z**-graded ring R and any integer n, the shift functor $T_n : R\text{-gr} \to R\text{-gr}$ is defined by $T_n(M) = M(n)$, where $M(n)$ is M as an ungraded R-module, but with the gradation defined by $M(n)_m = M_{n+m}$, for all $m \in \mathbf{Z}$.

II.1.22 Proposition Let R be a discrete gr-valuation ring; then $R(n_1) \oplus \cdots \oplus R(n_k) \cong_g R(m_1) \oplus \cdots \oplus R(m_k)$ for some positive integers n_i, m_j if and only if there exists a permutation σ on $\{1, \ldots, k\}$ such that $n_{\sigma(i)} = m_i$ modulo e, where e is the ramification index of R.

Proof: Let M_v be the gr-maximal ideal of R; then by the definition of e, we have $R(\underline{n})/MR(\underline{n}) = k[X^e, X^{-e}](\underline{n})$ and $R(\underline{m})/MR(\underline{m}) = k[X^e, X^{-e}](\underline{m})$, where for any graded R-module N, we define $N(\underline{m}) = \oplus_i N(m_i)$. Comparing $\dim_k((R(\underline{n})/MR(\underline{n}))_i)$ and $\dim_k((R(\underline{m})/MR(\underline{m}))_i)$ for every i yields the result. ■

If M and N are graded R-lattices over the **Z**-graded gr-Dedekind ring R, then we say that M and N have the same *genus* and we write $M \sim N$ if $M_p \cong N_p$ for all $P \in \mathrm{Max}^g(R)$. It is not hard to check the following properties. If $M \sim N$, then $f \in \mathrm{Hom}_{R\text{-gr}}(M, N)$ is an isomorphism if and only if $f_0 = f \mid_{M_0}$ is an isomorphism. If M and N are graded R-lattices, then $\det(M) = \det(N)$ if and only if $M_0 \cong N_0$. For graded fractional ideals I and J of R, such that $I \sim J$, we have $J = HI$ for some fractional R_0-ideal H. Using the approximation property for R_0, we may actually prove: if M, N are graded R-lattices, such that $M \sim N$ and $M \cong I_1 \oplus \cdots \oplus I_n$ for some graded fractional ideals $I_1, \ldots, I_n$ of R, then there exist $J_1, \ldots, J_n$ which are graded fractional ideals of R, such that $N \cong J_1 \oplus \cdots \oplus J_n$ and $I_i \sim J_i$ for $i = 1, \ldots, n$. As a further property of graded fractional ideals I, J of R, let us mention that

$HI \oplus KJ \cong I \oplus KHJ$, where H and K are fractional ideals of R_0.

II.1.23 Theorem Let M and N be graded R-lattices over the gr-Dedekind domain R. The following assertions are equivalent:
(1.23.1) $M \cong_g N$ in R-gr;
(1.23.2) $M \sim N$ and $M_0 \cong N_0$ in R_0-mod;
(1.23.3) $M \sim N$ and $\det(M) \cong \det(N)$;
(1.23.4) $M \sim N$ and $\det(M_0) \cong \det(N_0)$.

Proof: See [57]. ■

As a consequence of the above result, one may prove the cancellation property for gr-Dedekind domains:

II.1.24 Proposition If M, N, P are graded R-lattices over the gr-Dedekind domain R, then $M \oplus P \cong_g N \oplus P$ implies $M \cong_g N$. ■

Let us mention finally a rather useful property for gr-Dedekind rings satisfying GAP.

II.1.25 Proposition (1.25.1) Assume that R satisfies GAP and that $R^{(e)}$ is of the form $\tilde{R}_0(I)$ for some fractional R_0-ideal I, then $\mathrm{Cl}(R) = \mathrm{Cl}(R_0)/\langle[I]\rangle$.
(1.25.2) If R is gr-semilocal, then R is a gr-principal ideal domain if and only if R satisfies GAP. So, if a gr-semilocal gr-Dedekind ring satisfies GAP, then $\mathrm{Pic}^g(R) = 1$. (Also cf. III.3.2 and III.3.10.) ■

II.2 SEPARABILITY AND GRADED GALOIS EXTENSIONS

In this section, we will use the notation and definitions introduced by M. Raynaud in [54], chapter IV. Throughout this chapter, all gradations considered are **Z**-gradations. Let us start with the following result:

II.2.1 Proposition Let K be a graded field and S a (commutative) graded algebra of finite type; then for any graded prime ideal q of S, the following assertions are equivalent:
(2.1.1) q is isolated in $\mathrm{Spec}^g(S)$;
(2.1.2) $Q_q^g(S)$ is a finite graded K-algebra.

Proof: If q is isolated in $\mathrm{Spec}^g(S)$, then there exists $f \in h(S-q)$ such that $\mathrm{Spec}^g(S_f) = \{qS_f\}$. This S_f is gr-artinian and gr-local with gr-maximal ideal qS_f. Now, as S_f/qS_f is a graded field which is finite over K, it follows that S_f is finite over K. Moreover, as S_f is gr-local, $S_f = Q_q^g(S_f) = Q_q^g(S)$. Conversely, suppose that $Q_q^g(S)$ is a finite K-algebra. Consider the exact sequence

$$0 \longrightarrow N \longrightarrow S \longrightarrow Q_q^g(S) \longrightarrow M \longrightarrow 0.$$

Then $Q_q^g(N) = Q_q^g(M) = 0$ and N is a finitely generated S-module, as S is noetherian. Since $Q_q^g(S)$ is a finitely generated K-algebra, M is finitely generated as a K-module. It follows that there exists $f \in S - q$ such that $N_f = M_f = 0$, hence $Sf = Q_q^g(S)$. As $Q_q^g(S)$ is gr-local and finite over K, we obtain that $\{q\} = \mathrm{Spec}^g(Q_q^g(S)) = \mathrm{Spec}^g(S_f)$. ■

If R is a graded commutative ring, then for any $p \in \mathrm{Spec}^g(R)$, we write $k^g(p) = Q_p^g(R)/pQ_p^g(R)$ for the *graded residue class field of R at p*.

II.2.2 Proposition Let S be a commutative graded R-algebra of finite type and q a graded prime ideal of S lying over $p \in \mathrm{Spec}^g(R)$; then the following assertions are equivalent:
(2.2.1) q is isolated in $\mathrm{Spec}^g(S \otimes_R k^g(p))$;
(2.2.2) $Q_q^g(S)/pQ_q^g(S)$ is finite over $k^g(p)$.

Proof: We may suppose that R is gr-local with gr-maximal ideal p and then $S \otimes_R k^g(p) = S/pS$. Now, $Q_q^g(S)/pQ_q^g(S) = Q_q^g(S/pS)$ and we may apply (II.2.1) to the graded $k^g(p)$-algebra S/pS. ■

If S satisfies one of the equivalent conditions of (II.2.2), then S is said to be *gr-quasifinite over R at q*. We call S *gr-quasifinite over R* if it has this property at all $q \in \mathrm{Spec}^g(S)$. Mimicking the proof of [54], Proposition IV.3. one may easily verify that S is a gr-quasifinite R-algebra if and only if $S \otimes_R k^g(p)$ is a finite $k^g(p)$-algebra, for all $p \in \mathrm{Spec}^g(R)$.

We omit the proof of the following theorem, which is a rather straightforward adaptation of that of [54], Theorem IV.1.

II.2.3 Theorem (Graded Version of Zariski's Main Theorem) Let S be a graded R-algebra of finite type and let R' be the integral closure of R in S. If S is gr-quasifinite at $q \in \mathrm{Spec}^g(S)$, then there exists $f \in h(R' - q)$ such that $R'_f \cong_g S_f$. ■

II.2.4 Corollary Let S be a graded R-algebra of finite type; then S is a gr-quasifinite R-algebra if and only if there exists a finite graded R-algebra R' such that we have a factorization $R \to R' \to S$, and such that there exists $F = \{f_1, \dots, f_n\} \subset h(R')$, where $\{D^g(f_i) : i = 1, \dots, n\}$ covers $\mathrm{Spec}^g(S)$ and $R'_f \cong S_f$ for all $f \in F$ (in other words, $R' \to S$ is what one might call a *graded open immersion*).

Proof: One implication is straightforward, and the other follows from (II.2.3). ■

II.2.5 Corollary Let S be a graded algebra of finite type; then S is a gr-quasifinite R-algebra if and only if S is a quasifinite algebra in the sense of [54].

Proof: Compare (II.2.4) and [54], Corollary IV.2. ∎

The above results will be used below. Let us now introduce some graded analogs of the notions of étale and Galois extensions. Let us start with the following characterization.

II.2.6 Proposition Let S be a commutative graded R-algebra; then the following assertions are equivalent:
(2.6.1) S is separable R-algebra in the sense of [23, 37, ...], that is, the map $S \otimes_R S \to S$ splits.
(2.6.2) For each graded R-algebra T and for each graded ideal J of T with $J^2 = 0$, the canonical map $\mathrm{Hom}_{\text{gr-}R\text{-alg}}(S,T) \to \mathrm{Hom}_{\text{gr-}R\text{-alg}}(S,T/J)$ is injective.

Proof: (M. Raynaud introduced in [54] the notion of *algébre formellement nette*. By [54] Theorem III.2. and [54] Theorem III.1.4(g), it follows that this is the same as a separable algebra. Proposition (II.2.6) tells us that a graded separable algebra is just what we should call *une algébre gr-formellement nette.*). Let S be a separable R-algebra and put $I = \mathrm{Ker}(S \otimes S \to S)$ (resp. $\Omega(S/R) = I/I^2$). Then, by the results just mentioned, $I/I^2 = 0$ and S is an *R-algébre formellement nette*. So, from [54], Définition I.3, it follows that

$$\mathrm{Hom}_{R\text{-alg}}(S,T) \longrightarrow \mathrm{Hom}_{R\text{-alg}}(S,T/J)$$

is injective. This implies the result, if we restrict to R-algebra homomorphisms of degree zero.

Conversely, consider the graded R-algebra $T = S \oplus \Omega(S/R)$; then $\Omega(S/R)$ is a graded ideal of T and $\Omega(S/R)^2 = 0$. The identity map $S \to T/\Omega(S/R)$ may be lifted to: $u : S \to S \oplus R(S/R)$. Now, by [54], Theorem III.1., the set of maps u corresponds to $\mathrm{Hom}_S(\Omega(S/R), \Omega(S/R)) = \mathrm{HOM}_S(\Omega(S/R),\Omega(S/R))$, hence the set of graded liftings corresponds to $\mathrm{Hom}_{S\text{-gr}}(\Omega(S/R),\Omega(S/R))$. By assumption, u is the only graded lifting of the identity, so

$\mathrm{Hom}_{S\text{-gr}}(\Omega(S/R),\Omega(S/R)) = \{id\}$, hence $\Omega(S/R) = 0$ and S is a separable R-algebra. ∎

Before we give the definition of a gr-étale extension, let us first study the graded separable extensions of a graded field.

II.2.7 Proposition Let $L = l[X,X^{-1}]$ be a finite graded field extension of the graded field $K = k[T,T^{-1}]$, where $p = \mathrm{char}(k)$. Then the following statements are equivalent:
(2.7.1) L is a graded separable extension of K.
(2.7.2) For each K-linearly independent subset $\{x_1,\ldots,x_n\}$ of $h(L)$, the set $\{x_1^p,\ldots,x_n^p\}$ is K-linearly independent.
(2.7.3) There exists a homogeneous basis $\{y_1,\ldots,y_m\}$ of L over K such that $\{y_1^p,\ldots,y_m^p\}$ is a homogeneous basis of L over K.
(2.7.4) For each $x \in h(L)$, the minimal homogeneous polynomial of x over K has no multiple roots.
(2.7.5) l is a separable field extension of k, and $\deg T/\deg X$ is not a multiple of p.

Proof: The equivalence of (2), (3), and (4) is a direct translation of the corresponding classical theory of separable field extensions. Also note that one can define the minimal homogeneous polynomial of $x \in h(L)$ as a homogeneous generating element of the graded ideal $\{f \in K[X]; f(x) = 0\}$, where $\deg X = \deg x$. Also observe that $K[X]$ is a graded principal ideal domain.

For the implication (4) $\Rightarrow$ (5), observe that the minimal homogeneous polynomial of $x \in L_0 = l$ lies in $k[X]$, which implies that l/k is separable. Suppose $\deg T = pn \deg X$; then it is easily seen that the minimal homogeneous polynomial of X is of the form $f(X) = g(X^{pn})$. Then f has multiple roots, since $Df = 0$. If $x,y \in h(L_0)$ satisfy (4), then so does xy. Since $x \in l$ satisfies (4), it suffices to prove that x does. Let $d = \deg T = r \deg x$; then the minimal homogeneous polynomial of X can be chosen of the form

$f(X) = a_0 + a_1X^r + a_2X^{2r} + \cdots + a_nX^{nr}$, where $\deg a_i = -id$. Since r is no multiple of p, $Df = 0$, so X is a simple root. This proves that (5) $\Rightarrow$ (4).

To prove that (4) $\Rightarrow$ (1), it suffices to consider extensions of the form $L = K[X]/(f(X))$, where f is a homogeneous irreducible polynomial (by [54], III.2.4, if $K \subset K_1 \subset L$ are graded fields, then L/K is separable if and only if L/K_1 and K_1/K are separable). Let M be a finite graded extension of K decomposing $f : f(X) = \prod_{i \in I}(X - a_i)$, for $a_i \neq a_j \in h(M)$; then we have a graded isomorphism of $L \otimes M = M[X]/(f(X))$ into a finite product $\prod_{i \in I} M[X]/(X - a_i)$, hence $L \otimes M/M$ is separable, and so is L/K ([54], III.2.2).

Finally, to prove (1) $\Rightarrow$ (4), suppose (4) is not true; then we may find $x \in h(L)$ such that the derivative of the minimal homogeneous polynomial vanishes. But then $D = d/dx$ is a K-derivation, so L/K is not separable. ■

It follows easily from (II.2.7) that every graded separable extension of finite type of a graded field K is of the form $\prod_{i \leq n} L_i$, where L_i is a separable graded field extension of K.

Let S be a commutative graded R-algebra. We say that S is a *gr-étale R-algebra* if S is finitely presented as an R-algebra and if for each graded R-algebra T and each graded ideal J of T with $J^2 = 0$, we have

$$\mathrm{Hom}_{\text{gr-}R\text{-alg}}(S,T) \cong \mathrm{Hom}_{\text{gr-}R\text{-alg}}(S,T/J).$$

As in [54], Chapter 2, it is very easy to verify the following properties:

II.2.8 Lemma Let R be a graded commutative ring, let R' and S be commutative graded R-algebras, and T a commutative graded S-algebra; then the following results hold:

(2.8.1) If S is gr-étale over R and T is gr-étale over S, then T is gr-étale over R.
(2.8.2) If R' and S are gr-étale over R, then so is $R' \otimes_R S$.
(2.8.3) If R' is a faithfully flat R-algebra and $R' \otimes_R S$ is a gr-étale R'-algebra, then S is a gr-étale R-algebra.
(2.8.4) If for all $q \in \mathrm{Spec}^g(S)$ there exists $f \in h(S-q)$, such that S_f is gr-étale over R, then so is S.
(2.8.5) If f and g are homogeneous polynomials in $R[X, \deg X = t]$, and if f' is invertible in $S = (R[X]/(X))_g$, then S is a gr-étale R-algebra.

We call this a *standard gr-étale* R-algebra.

Let S be a commutative R-algebra and $q \in \mathrm{Spec}^g(S)$, then we say that S is *graded separable of finite type* (resp. *gr-étale*) *in a neighborhood of* q if there exists $f \in h(S-q)$ such that S_f is graded separable of finite type (resp. gr-étale) over R.

II.2.9 Theorem Let S be a commutative graded R-algebra and q a graded prime ideal of S lying over p. Then
(2.9.1) S is gr-étale in a neighborhood of q if and only if there exists $f \in h(S-q)$ and $h(R-p)$ such that S_f is graded isomorphic to a standard gr-étale R_h-algebra, that is, $S_f \cong_g (R_h[X]/(p))_g$.
(2.9.2) S is graded separable of finite type in a neighborhood of q if and only if there exists $f \in h(S-q)$, $h \in h(R-p)$, a standard gr-étale R_h-algebra C, and a surjective graded morphism $u : C \to S_f$.

Moreover, in this last case the map $u \otimes_R k^g(p) : C \otimes_R k^g(p) \to S_f \otimes_R k^g(p)$ is a graded isomorphism.

Proof: In both cases, one implication is obvious, using (2.8.5) and the fact that the quotient of an étale extension is separable of finite type. The proof of the converse statements is more difficult, but rather similar to the proof of the corresponding ungraded result. We will restrict ourselves to a brief sketch of the proof.

Step 1 By a standard argument, we can reduce the proof to the case where R is gr-local with gr-maximal ideal p.

Step 2 Using the graded version of Zariski's main theorem, we can reduce the proof to the case where S is a finite R-algebra. For instance, suppose S is graded separable of finite type; then S is quasifinite by [54], Chapter IV, hence gr-quasifinite. So, by (II.2.4), there exists a finite graded subalgebra T of S, finite over R, $g \in h(T)$, and $f \in h(S-q)$ such that $S_f \cong_g T_h$. It now suffices to replace S by T. The gr-étale case is done in a similar way.

Step 3 (Reduction to the case $S = R[x]$, x homogeneous) Let $K = R/p$ and $S' = S \otimes K$; then S'_q is a separable graded field extension of K, so we can find a homogeneous $x' \in S'$ such that the image of x' in $S'_{q'}$ generates $S'_{q'}$ over K and such that the images of x' in the other components of S' are 0. Lift x' to a homogeneous $x \in S$ and let $C = R[x] \subset S$. Let $r = q \cap C$; then an application of the graded version of Nakayama's lemma yields that $Q^g_r(C) = Q^g_q(S)$. Now, as C and S are finite over R, there exists $f \in h(C-r)$ such that $S_f \cong C_f$. So we may replace S by C.

Step 4 (Proof of (2.9.2)) Let S have rank r; then S is generated over R by $1, x, \ldots, x^{r-1}$, so there exists a monic polynomial $P \in R[X, \deg X = \deg x]$ of degree r such that $P(x) = 0$. Hence S is a quotient of $R[X]/P$, and the induced map $K[X]/(P') \to S'$ is an isomorphism. Let q'' be the inverse image of q in $R[X]/(P)$; then $R[X]/(P)$ is graded separable over R at q'', so the image of $P' = (d/dx)P$ is invertible in $C = (R[X]/(P))_g$, which is gr-étale of standard type, and there exists $h \in B - q$ such that B_h is a quotient of C.

Step 5 (Proof of (2.9.1)) Let S be gr-étale over R; then by step 4, we already have a standard R-algebra C and an exact sequence

$$O \longrightarrow I \longrightarrow C \xrightarrow{u} S \longrightarrow 0$$

where $u' : C' \to S'$ is an isomorphism.

Let $r = u^{-1}(q)$; then we will show that $I = 0$ in a neighborhood of r. Clearly I is of finite type, since C and S are finitely presented, and it is sufficient to show that $Q_r^g(I) = 0$. By the graded version of Nakayama's lemma, we therefore need only to show that $Q_r^g(I)/Q_r^g(I)^2 \cong Q_r^g(I/I^2) \cong 0$.

Consider the exact sequence

$$(*) \quad 0 \longrightarrow I/I^2 \longrightarrow C/I^2 \longrightarrow S/I^2 \longrightarrow 0;$$

then I/I^2 is a graded ideal of C/I^2 and $(I/I^2)^2 = 0$. As S is gr-étale, the isomorphism $S \cong (C/I^2) \cong C/I$ lifts to a graded morphism $S \to C/I^2$, so $(*)$ splits as an exact sequence of graded modules. Tensoring with K, we still have an exact sequence

$$0 \longrightarrow (I/I^2) \otimes_R K \longrightarrow (C/I^2) \otimes_R K \longrightarrow S' \longrightarrow 0,$$

where the surjective map is an isomorphism, since u' is, hence $(I/I^2) \otimes_R K = 0$. By the graded version of Nakayama's lemma, $Q_r^g(I/I^2) = 0$. ∎

II.2.10 Corollary Let S be a commutative graded R-algebra; then the following assertions are equivalent:

(2.10.1) S is a gr-étale R-algebra;

(2.10.2) S is an étale R-algebra;

(2.10.3) S is separable and finitely presented as an R-algebra and flat as an R-module.

Proof: The equivalence of (2) and (3) is just Corollary 2 to Theorem V.2 in [54]. Suppose that S is gr-étale; then Theorem (II.2.9) implies that S is étale of standard type at any point of $\mathrm{Spec}^g(S)$. As $\mathrm{Spec}^g(S)$ is dense in $\mathrm{Spec}(S)$, it follows that S is étale over R. Finally, for the implication (3) $\Rightarrow$ (1), we may suppose that R is gr-local with maximal ideal p. Let q be a graded prime ideal of S lying over p. We may suppose that there exists a gr-étale R-algebra of standard type T and a surjective graded morphism $u : T \to S$ such that the induced map u' is an isomorphism. Let

$I = \text{Ker}(u)$; then i is of finite type, as S and T are finitely presented. Let r be the inverse image of q in T; then we will show that $I = 0$ in a neighborhood of r, or $Q_r^g(I) = 0$. Since $Q_q^g(S)$ is flat as an R-algebra, we have an exact sequence

$$0 \longrightarrow (Q_r^g(I))' \longrightarrow (Q_r^g(T))' \longrightarrow (Q_r^g(S))' \longrightarrow 0.$$

From $T' = S'$, it follows that $(Q_r^g(T))' \cong (Q_r^g(S))'$, so $(Q_r^g(I))' = 0$ and $Q_r^g(I) = 0$, by the graded version of Nakayama's lemma. ■

The following straightforward result will be used below:

II.2.11 Proposition The idempotents of a commutative graded ring are homogeneous of degree zero.

Proof: Suppose first that R is a reduced gr-local ring; then, if e is a nonhomogeneous idempotent, the component of highest or lowest degree is a nilpotent element, which contradicts the hypothesis. We therefore have that 0 and 1 are the only idempotents in R.

Next, assume that R is an arbitrary gr-local ring, and let m be its gr-maximal ideal; then $R' = R/m$ is a graded field, so if e is an idempotent in R, then $e' = 0$ and $e \in m$. Let N be the nilradical of R. From the reduced case, we know that $e = 0 \bmod N$ in R/N, hence $e \in N$. But then e is a nilpotent idempotent, so $e = 0$. If $e' = 1$, then $1 - e' = 0$, so $e = 1$.

In general, suppose e is an idempotent of R, and let e_0 be the part of degree zero in the homogeneous decomposition. For every graded prime ideal p, the element $e - e_0$ maps to 0 in $Q_p^g(R)$, hence $e - e_0 = 0$. ■

Let S be a commutative graded ring extension of R, and G a finite group of R-automorphisms of S. We introduce the following notation: $\Delta(S : G)$ is the graded R-algebra which has a free basis $\{u_\sigma : \sigma \in G\}$ as a graded S-module, and multiplication rules

$(au_\sigma)(bu_\tau) = a\sigma(b)u_{\sigma\tau}$. We have a natural R-algebra homomorphism $j : \Delta(S : G) \to \mathrm{End}_R(S)$, given by $j(au_\sigma)(x) = a\sigma(x)$.

$\nabla(S : G)$ is the direct sum of $|G|$ copies of S, with free basis $\{v_\sigma : \sigma \in G\}$ as a graded S-module. We have an R-algebra homomorphism $l : S \otimes S \to \nabla(S : G)$ given by $l(a \otimes b) = \sum_{\sigma \in G} a\sigma(b)v_\sigma$, which is an S-module homomorphism if we consider $S \otimes S$ as an S-module, by letting S act on the first factor.

II.2.12 Theorem Let S be a graded R-algebra which is a Galois extension of R with group G; then any R-automorphism of S is graded.

Proof: Consider the R-algebra isomorphism $l : S \otimes_R S \to \nabla(S : G)$ introduced before, and let $e_\sigma = l^{-1}(v_\sigma)$; then e_σ is an idempotent, so it has degree zero. It follows that l^{-1}, which is given by $l^{-1}(\sum_\sigma a_\sigma l_\sigma) = \sum a_\sigma e_\sigma$, has degree zero, hence l has degree zero. One uses the fact that l is an S-module isomorphism, if one defines an S-module structure on $S \otimes_R S$ by letting S act on the first factor. Next, consider $b \in S_r$ and let $p_\sigma : \nabla(S : G) \to S$ be the projection on the component Sv_σ; then $(p_\sigma \circ l)(l \otimes b) = \sigma(b)$ has degree r, so $\deg \sigma = 0$ and σ is graded, indeed. ∎

If S satisfies the conditions of the previous theorem, then we call S a *graded Galois extension* or *gr-Galois extension* of R.

II.2.13 Corollary Let S be a gr-Galois extension of R; then the canonical isomorphisms $j : \Delta(S : G) \to \mathrm{END}_R(S)$ and $l_n : S^{(n)} \to K^n(G, S)$ are graded.

Proof: The statement follows from the fact that $\deg \sigma = 0$, and by observing the definitions of the morphisms j and l_n : $j(au_\sigma)(x) = a\sigma(x)$ and $l_n(a_1 \otimes \cdots \otimes a_{n+1}) = a_1\sigma_1(a_2) \cdots \sigma_{n-1}(a_n)\sigma_n(a_{n+1})$. ∎

II.2.14 Corollary Let S be a graded ring extension of R and let G be a finite group of R-automorphisms of S; then S is a gr-Galois extension of R with Galois group G if and only if
(2.14.1) $G = S^G$.
(2.14.2) For all $\sigma \in G$, $\deg \sigma = 0$.
(2.14.3) For each gr-maximal ideal m of S and for each $\sigma \neq 1$ in G there exists $x \in S$ such that $\sigma(x) - x \notin m$.

Proof: One implication is clear. Conversely, suppose that (1), (2), and (3) are satisfied; then for $\sigma \neq 1$, consider the graded ideal I of S generated by elements of the form $\sum_{i \leq n} x_i(y_i - \sigma(y_i))$, where $x_i, y_i \in h(S)$. Since I cannot be in any gr-maximal ideal, $I = S$. The rest of the proof is as in the nongraded case (cf. [23]). ■

II.2.15 Proposition (Embedding Theorem) Let S be a graded extension of R which is finitely generated projective and separable over R, and which has no nontrivial idempotents; then there exists a graded extension N of S which is a gr-Galois extension of R and which has no nontrivial idempotents.

Proof: A rather easy, but long modification of the corresponding ungraded result (cf. [23], Theorem III.2.9). ■

II.2.16 Corollary Let S be as in the preceding proposition; then every separable R-subalgebra T of S is a graded subalgebra.

Proof: Consider the gr-Galois extension N of R obtained above, By the fundamental theorem of Galois theory, there is a subgroup H of Gal(N/R) such that $T = N^H$. Writing $x = x_{i(1)} + \cdots + x_{i(n)}$ for the homogeneous decomposition of $x \in T$, we obtain $\sigma(x) = \sigma(x_{i(1)}) + \cdots + \sigma(x_{i(n)})$. From the uniqueness of the homogeneous decomposition and the fact that $\deg \sigma = 0$, it follows that $\sigma(x_{i(j)}) = x_{i(j)}$ for all $\sigma \in G$. Therefore $x_{i(j)} \in T$ and T is graded. ■

A graded field K is called *gr-separably closed* if it does not admit any nontrivial finite separable graded field extension. We say that K is *gr-algebraically closed* if F does not admit any nontrivial graded field extension which is algebraic over it.

Let us conclude this section with the following characterization:

II.2.17 Proposition (2.17.1) A gr-separably closed field K is of the following type:
(a) $K = k$, where k is separably closed, or
(b) $K = k[T,T^{-1}]$, where k is separably closed and $\text{char}(k) = 0$, $\deg T = 1$, or $\text{char}(k) = p$, $\deg T = p^n$.
(2.17.2) A gr-algebraically closed field is of the form k or $k[T,T^{-1}]$, where k is algebraically closed and $\deg T = 1$.

Proof: Left to the reader. ∎

It follows easily that for any graded field K, there exist embeddings $K \to K^s \to K^a = \tilde{K}$, where K^s and K^a are the solutions of some universal problem, K^a is gr-algebraically closed, K^s is gr-separably closed, and where K^s and K^a are unique up to graded isomorphism.

The only gr-étale extensions of K^s are of the form $K^s \otimes \cdots \otimes K^s$.

II.3 GRADED COMPLETION AND HENSELIZATION

In this section, R will denote a graded ring and I a (two-sided) ideal of R. The ideal I defines the I-adic valuation on R, given by $v(0) = \infty$ and $v(x) = \max\{n \in \mathbf{N};\ x \in I^n\}$ for $x \neq 0$. Choose a fixed $\alpha \in]0,1[$ and put $d(x,y) = \alpha^{v(x-y)}$; then (R,d) is an ultrametric space, which is a Hausdorff space if and only if

$\bigcap_n I^n = (0)$. This condition is satisfied, for example, when R is commutative and noetherian. In the sequel, we will always assume it to hold. The zero-dimensional topology $\mathcal{T}$-I generated by d is called the *I-adic topology* on R. The I_0-adic valuation on R_0 will be denoted by w. We say that R is *I-adically complete* if it is complete for the ultrametric d. If R is not complete, one may construct the completion $\hat{R}$, which is not a graded ring in general, however (e.g., take $R = k[X]$, $I = (X)$; then $\hat{R} = k[[X]]$).

We therefore introduce the notion of gr-complete rings. For each $n \in \mathbf{N}_0$ we know that I^n is a graded ideal, so $I^n = \oplus_{r \in Z}(I^n)_r$. For each $r \in \mathbf{Z}$, let v_r be the restriction of v to R_r; then, for $x_r \neq 0$ in R_r, we have $v_r(x_r) = \max\{n \in \mathbf{N} : x_r \in (I^n)_r\}$ and for $x \in R$ we have $v(x) = \min\{v_r(x_r) : r \in \mathbf{Z}\}$. Also note that in general $v_0(x) \neq w(x)$.

We say that the graded ring R is *gr-I-complete* if R_r is complete for the valuation v_r, for each $r \in \mathbf{Z}$. If R is not gr-I-complete, we may consider the completion R_r of $\hat{R}_r$ with respect to v_r. Clearly v_r extends to $\hat{R}_r$. Moreover $\hat{R}_0$ is a ring and $\hat{R}_r$ is an $\hat{R}_0$-module. Before we introduce the notion of a graded completion, let us establish the following lemma:

II.3.1 Lemma With notation as above we have:
(3.1.1) $\hat{R} = \{(x_r)_r \in \prod_{r \in \mathbf{Z}} \hat{R}_r;\ \lim_{|r| \to \infty} v_r(x_r) = \infty\}$;
(3.1.2) $\hat{R}_r \hat{R}_s \subset \hat{R}_{r+s}$.

Proof: Let $(x^{(n)})_n$ be a v-Cauchy sequence; then for each $r \in \mathbf{Z}$, $(x_r^{(n)})_n$ is a v_r-Cauchy sequence. So, putting $x_r = v_r - \lim x_r^{(n)}$ in $\hat{R}_r$, we get easily that $x = (x_r)_r = v - \lim x^{(n)} \in \prod \hat{R}_r$. Hence for all $M > 0$, there exists $N \in \mathbf{N}$ such that for all $n \geq N$ we have $v(x^{(n)} - x) > M$. As $x^{(N)} \in R$, there is $s > 0$ such that $x_r^{(N)} = 0$ for $|r| \geq s$. But then, for each $|r| \geq s$ we have $v_r(x_r) = v_r(x_r^{(N)} - x_r) > v(x^{(N)} - x) > M$, so $\lim v_r(x_r) = \infty$.

Conversely, take $x = (x_r)_r$ in $\prod_r \hat{R}_r$ such that $\lim_{|r|\to\infty} v_r(x_r) = \infty$ for each $r \in \mathbf{Z}$; then we have a sequence $(x_r^{(n)})_n$ in R_r such that $v_r - \lim x_r^{(n)} = x_r$. So for all $M > 0$ and $r \in \mathbf{Z}$, there exists $N_r(M)$ such that $v_r(x_r^{(n)} - x_r) > M$ for all $n \geq N_r(M)$. There also exists $R(M) > 0$ such that $v_r(x_r) > M$ for all $|r| > R(M)$. Now, let $N(M) = \max\{N_r(M) : |r| \leq R(M)\}$ and define $y^{(n)} \in R$ by:

$$y_r^{(n)} = \begin{cases} x_r^{(N(n))} & \text{for } |r| \leq R(n), \\ 0 & \text{for } |r| > R(n). \end{cases}$$

It is then easily checked that $v - \lim y^{(n)} = x$. The proof of the second part of the lemma is obvious. ■

It follows from the preceding lemma that $\hat{R}$ is a filtered ring. We define the *graded completion* $\hat{R}^g$ as the graded ring associated to $\hat{R}$. It is clear that we have $\hat{R}^g = \oplus_{r\in\mathbf{Z}} \hat{R}_r$. We will show below that $\hat{R}^g$ can also be defined by means of a nonarchimedean uniformity.

Let F be the set of functions from $\mathbf{Z}$ to the half-open interval $]0,1]$. For each σ in F and x in R, we define $v_\sigma(x) = \min\{v_r(x_r)\sigma(r) : r \in \mathbf{Z}\}$; then, fixing $\sigma \in]0,1]$, let $d_\sigma(x,y) = \alpha v_\sigma(x-y)$. This defines another ultrametric on R and we denote by U-I-gr the uniformity generated by $P = \{d_\sigma : \sigma \in F\}$. It is easily seen that $\mathcal{U}$-I-gr is a nonarchimedean uniformity (cf. [70], p. 34).

II.3.2 Theorem The completion of R with respect to the $\mathcal{U}$-I-gr-uniformity coincides with the graded completion $\hat{R}^g$ of R.

Before giving the proof of this result, let us first establish the following lemma:

II.3.3 Lemma $(x^{(n)})_n$ is a $\mathcal{U}$-I-gr-Cauchy sequence in R if and only if the following conditions are satisfied:
(3.3.1) For each $r \in \mathbf{Z}$, $(x_r^{(n)})_n$ is a v_r-Cauchy sequence in R_r.
(3.3.2) There exists a finite subset $J \subset \mathbf{Z}$ such that $x^{(n)} \in \oplus_{r \in J} R_r$, for all $n \in \mathbf{N}$.

Proof: Suppose (1) and (2) are satisfied. Choose $M > 0$ and $\sigma : \mathbf{Z} \to]0,1]$; then from (1) we get that for all $r \in J$ there is N_r such that for all $n, m > N_r$ we have $v_r(x_r^{(n)} - x_r^{(m)}) > M/\sigma(r)$. Set $N = \max\{N_r : r \in J\}$; then for all $n, m > N$ and all $r \in \mathbf{Z}$, we have $\sigma(r)v_r(x_r^{(n)} - x_r^{(m)}) > M$. Moreover, this is trivial if $r \notin J$. Conversely, let $(x^{(n)})_n$ be a $\mathcal{U}$-I-gr-Cauchy sequence; then (1) follows immediately. Suppose (2) is false; put $J = \{r \in \mathbf{Z};\ x_r^{(n)} \neq 0$ for some $n\}$ and assume $\sup J = +\infty$. The case $\inf J = -\infty$ is treated in a similar way. For $x \in R \subset \prod \hat{R}_r$, we introduce the following general notation: $g(x) = \sup\{r \in \mathbf{Z};\ x_r \neq 0\}$. Up to replacing $(x^{(n)})_n$ by a subsequence, we may suppose that $g(x^{(n)}) = m(n)$, with $m(1) < m(2) < m(3) < \cdots$. We define σ in F as follows:

$$\sigma(m(j)) = 1/v_j(x_{m(j)}^{(j)}) \qquad \text{if } v_j(x_{m(j)}^{(j)}) \neq 0;$$
$$\sigma(n) = 0 \qquad \text{in all other cases.}$$

Then $v_\sigma(x^{(k)} - x^{(j)}) \leq \sigma(m(k))v_k(x_{m(k)}^{(k)}) \leq 1$, so it follows that $(x^{(n)})_n$ is not a $\mathcal{U}$-I-gr-Cauchy sequence. ∎

Proof (of Theorem (II.3.2)): From the preceding results, we know that $\oplus_{r \in Z} \hat{R}_r$ may be embedded in the completion of R with respect to U-I-gr and that no $\mathcal{U}$-I-Cauchy sequence converges to an element of $\prod_r \hat{R}_r - \oplus_r \hat{R}_r$. Since in general, $\mathcal{U}$-I-gr does not have a countable basis, we still have to convince ourselves (and the reader!) that no Cauchy *net* converges to $x \in \prod_r \hat{R}_r - \oplus_r \hat{R}_r$. Suppose that $(x^{(\alpha)})_{\alpha \in x}$ is such a Cauchy net. We assume that $g(x) = +\infty$, as in the lemma above, and that $(x^{(\alpha)}) \to x$ in the $\mathcal{U}$-I-gr-topology. Recall that each cofinal subnet $(x^{(\beta)})_{\beta \in Y}$ also converges to x, a subnet being called *cofinal* if for all $\alpha \in X$ there exists $\beta \in Y$ such that $\beta \geq \alpha$.

Define X_i as follows ($i = 1, 2, \ldots$);

$$X_1 = \{\alpha : g(a^{(\alpha)}) \geq 1\};$$

$$\vdots$$

$$X_{n+1} = X_n - \{\alpha : g(x^{(\alpha)}) = n+1,\ g(x^{(\beta)}) \leq n,$$

$$\text{for some } \beta \geq \alpha\}.$$

Let $Y = \cap X_n$; then $(x^{(\beta)})_{\beta \in Y}$ has the property that for all $\alpha, \beta \in Y$, $\alpha \leq \beta$ implies $g(x^{(\alpha)}) \leq g(x^{(\beta)})$.

We also have that $g(x^{(\alpha)}) \leq n$ and $\alpha \in X_n$ imply $\alpha \in Y$. Next, for all $n \in N$, there is $m \geq n$ and $\alpha \in X_m$ such that $g(x^{(\alpha)}) = m$. Indeed, otherwise we have for all α that $g(x^{(\alpha)}) \geq n$ implies that there exists $\beta \geq \alpha$ with $g(x^{(\beta)}) < n$. Then $x^{(\alpha)} \to x$ implies that $g(x) \leq n$, which is a contradiction. This implies that we may take $x^{(\alpha_1)}, x^{(\alpha_2)}$ such that $g(x^{(\alpha_i)}) = m_i$ and $x^{(\alpha_i)} \in X_{m_i}$, $m_1 < m_2 < \cdots$ It follows that $(x^{(\beta)})_{\beta \in Y}$ is a cofinal subset. Indeed, take $\alpha \in X - Y$ and suppose that $g(x^{(\alpha)}) = n$; then there is a $\beta \in X_{n-1}$ such that $g(x^{(\beta)}) \leq n - 1$ and $\beta \geq \alpha$. But then $\beta \in Y$.

Finally, $(x^{(\alpha_i)})_{i \in \mathbf{N}}$ is a cofinal subsequence of $(x^{(\beta)})_{\beta \in Y}$, for if it were not, there would exist $\beta \in Y$ such that $\beta > \alpha_i$ for each $i \in \mathbf{N}$, so $g(x^{(\alpha)}) > x^{(\alpha_i)} = m_i$, so $g(x^{(\beta)}) = +\infty$, which is a contradiction. Now $(x^{(\alpha_i)})_{i \in \mathbf{N}}$ is a sequence converging to x, so from the foregoing lemma, it follows that $g(x) < \infty$, establishing the result. ■

II.3.4 Proposition If for each $N \in \mathbf{N}$ there exists n with $|n| > N$ such that for all $M \in \mathbf{N}$, there is $x \in R^n$ such that $v(x) > M$, then $\mathcal{U}$-I-gr is not metrizable.

Proof: If $\mathcal{U}$-I-gr were metrizable, then it would have a countable basis, hence one would be able to generate it by a countable subset $P = \{d_\sigma : \sigma \in F\}$.

Let $\mathcal{U}$ be the uniformity generated by some $\{d_{\sigma(i)} : i \in \mathbf{N}\} \subset P$; we shall construct a $\mathcal{U}$-Cauchy sequence which is not a $\mathcal{U}$-I-gr-Cauchy sequence. For each $r \in \mathbf{Z}$, define $p(r) = \exp\sup\{1/\sigma_i(r) : i \leq r\}$; then we have that $\lim_{r\to\infty} \sigma_i(r) = 0$, because, in other situations, $d_{\sigma(i)}$ becomes equivalent to d and the statement is trivial. Using induction, one can construct a sequence $(x_i)_i$ in $h(R)$ such that $\deg x_i = n(i)$, with $|n(1)| < |n(2)| < \cdots$ and $v_{n(i)}(x_i) > p(n(i))$. Start, for example, with $n(0) = 0$; then, putting $N = |n(i)|$, we get $|n(i)| > |n(i-1)|$ such that there exists x_i with $\deg x_i = n(i)$ and $v_{n(i)} > p(n(i))$.

Let $y_n = \sum_{i\leq n} x_i$; then $(y_n)_n$ is not a $\mathcal{U}$-I-gr-Cauchy sequence (from the last lemma), but it is a $\mathcal{U}$-Cauchy sequence, as we have for each $i \in \mathbf{N}$ and $t > s$ that $v_{\sigma(i)}(y_t - y_s) = v_{\sigma(i)} = \min_{s<j\leq t} \sigma_i(n(j)) v_{n(j)}(x_j) > \min_{s<j\leq t} \sigma_i(n(j)) p(x_{n(j)})$, which goes to $+\infty$ as s and t do. ■

II.3.5 Examples Let $R = \mathbf{Z}[X]$, $\deg X = 1$ and $I = (p, X)$ for some prime p; then $\mathcal{U}$-I-gr is not metrizable. We find $\hat{R}^g = \mathbf{Z}_p[X]$ and $\hat{R} = \mathbf{Z}_p[[X]]$. On the other hand, let $R = k[X]$, with $\deg X = 1$ and $I = (X)$, where k is a field; then $\hat{R}^g = k[X]$ and $\hat{R} = k[[X]]$. Note that in this case $\mathcal{U}$-I-gr is metrizable: We have $v_r(X^r) = r$. Put $\sigma(0) = 1$ and $\sigma(0) = 1/r$ for $r \neq 0$; then v_σ generates $\mathcal{U}$-I-gr, since it generates the discrete topology, as $v_\sigma(x, y) < 1$ if $x \neq y$.

II.3.6 An Algebraic Characterization of gr-Completeness Using the fact that R is gr-I-complete if and only if every R_r is complete for v_r, it is easily verified that for each integer $r \in \mathbf{Z}$, the canonical map $\nu_r : R_r \to \varprojlim R_r/(I^n)_r$ is a bijection. The injectivity of this map is equivalent to $\cap I^n = (0)$, its surjectivity to one of the following properties:

(1) Given $a_1, a_2, \ldots$ in R_r such that $a_{i+1} - a_i \in (I^i)_r$, there exists $a \in R_r$ such that $a - a_i \in (I^i)_r$, for each i.
(2) Given a finite subset $E \subset \mathbf{Z}$ and $a_1, a_2, \ldots \in \oplus_{r \in E} R_r$ such that $a_{i+1} - a_i \in I^i$ for all $i \in \mathbf{N}$, there exists $a \in R$ such that $a - a_i \in I^i$ for all $i \in \mathbf{N}$.

Suppose that R is gr-I-complete; then we would like to know whether R_0, the part of degree 0, is I_0-complete. In the next proposition, we show that in the commutative case at least, this is true in general.

II.3.7 Proposition If the graded commutative ring R is gr-I-complete, then R_0 is I_0-complete (always under the assumption that $\cap I^n = (0)$!).

Proof: Since $(I_0)^n \subset (I^n)_0$, we have $w(x) \leq v_0(x)$, for all x in R_0 (but not necessarily the converse). Now, consider a w-Cauchy sequence $(x_n)_n$ in R_0; then $(x_n)_n$ has a limit x in the w-completion S_0 of R_0. Let $S = S_0 \otimes R$ and $J = S_0 \otimes I$; then v extends to a valuation on S and v_0 extends to a valuation on S_0. So (S_0, v_0) becomes a Hausdorff space. As $(x_n)_n$ is also a v-Cauchy sequence, it has a v_0-limit in R_0, say y. But since $(x_n)_n$ converges to x in the w-topology, it converges to x in the v_0-topology. It follows that $x = y$, hence $R_0 = S_0$. ∎

II.3.8 Note It is not necessarily true that the restriction of the $\mathcal{U}$-I-gr uniformity to R_0 is equal to the I_0-uniformity. As an example, suppose $R = k[Y, X_1, X_2, \ldots]$ with $\deg X_i = i$, $\deg Y = 1$, and $I = (Y, X_1, X_2, \ldots)$; then $w(Y^{m-1}X_{m-1}) = 2m$. So $Y^{m-1}X_{m-1}$ converges to 0 in the $\mathcal{U}$-I-gr-topology, but not in the I_0-adic topology. It will follow from II.3.11 below, that the statement is true if R contains a homogeneous unit with degree different from zero in its center. We first need a simple arithmetical lemma.

II.3.9 Lemma Let $e \neq 0$ be a positive integer and let $z_1, \ldots, z_e \in \mathbf{Z}$; then there exists $E \subset \{1, \ldots, e\}$ such that $\sum_{j \in E} z_j = 0 \bmod e$.

Proof: If for each subset J of $\{1, \ldots, e\}$ we had $\sum_{j \in E} z_j \neq 0 \bmod e$, then $z_1 \bmod e$, $(z_1 + z_2) \bmod e$, ..., $(z_1 + \cdots + z_e) \bmod e$ would be e different elements of $\mathbf{Z}/e\mathbf{Z}$, so one of them would have to be zero. ■

II.3.10 Corollary Let e, n be nonzero positive integers. If $z_1, \ldots, z_{ne}$ are integers such that $\sum_{i \leq ne} z_i = 0 \bmod e$, then there exists a partition of the set $\{1, \ldots, ne\}$, say $I(1) \cup \ldots U I(n)$, such that for $i = 1, \ldots, n$ we have $\sum_{j \in I(i)} z_j = 0 \bmod e$. ■

II.3.11 Proposition Consider a graded ring R containing an invertible homogeneous element of nonzero degree in its center; then for every graded ideal I of R, the I_0-adic uniformity on R_0 and the restriction to R_0 of the gr-I-adic uniformity on R coincide.

Proof: Up to replacing x by x^{-1}, we may assume that $\deg x = e > 0$. Obviously, $(I_0)^n \subset (I^n)_0$ for all $n \in \mathbf{N}$; hence it will suffice to establish that for all $n \in \mathbf{N}$, there exists $m \in \mathbf{N}$ such that $(I^m)_0 \subset (I_0)^n$. Put $m = en$. If $y \in (I^m)_0$, then $y = \sum r(\underline{a}) a_1 a_2 \cdots a_m$, with $r(\underline{a}) \in R_0$ and $a_i \in I$ homogeneous, such that $\sum_{j \leq m} \deg a_i = 0$.

Applying (II.3.10) with $z_i = \deg a_i$, $i = 1, \ldots, m$ we obtain n sets $I(1), \ldots, I(n)$ forming a partition of $\{1, \ldots, m\}$ such that $\sum_{i \in I(j)} z_i = w_j e$ with $w_j \in \mathbf{Z}$. Now $a_1 a_2 \cdots a_m = \prod_{j \leq n} (\prod_{i \in I(j)} a_i) = \prod_{j \leq n} (x^{-w_j} \prod_{i \in I(j)} a_i) \in (I_0)^n$, because $\sum_{j \leq n} w_j = 0$ and x is central. ■

II.3.12 Notes The converse of Proposition (II.3.7) does not hold in general. Indeed, let $S = S_0 \otimes R$, where S_0 is the w-completion of R_0, and R is as in (II.3.6); then

$(\sum_{m\leq n} Y^{m-1}X_{m-1})_n$ is a Cauchy sequence in $\mathcal{U}$-I-gr having *no* limit. Note also that the v_0-completeness of R_0 does not imply the gr-completeness of R: let $S = S' \otimes R$, where R and R_0 are again as before and S' is the v_0-completion of R_0; then $(\sum_{m\leq n} Y^{m-1}X_{m-1})_n$ is a $\mathcal{U}$-I-gr-Cauchy sequence, having *no* limit.

It is well known that complete local rings satisfy Hensel's lemma (cf. [46, 54]), that is, they are henselian rings. An advantage of this notion is that it can also be applied to the non-noetherian case. Let us briefly discuss the notion of gr-henselian rings.

We call a commutative gr-local ring R *gr-henselian* if every commutative finite graded R-algebra is *gr-decomposed*, that is, if it is the direct sum of gr-local rings. Using the obvious fact that every finite graded algebra over a graded field is gr-artinian, we have:

II.3.13 Proposition If R is gr-local ring with gr-maximal ideal m, then every finite graded R-algebra B is gr-semilocal and the gr-maximal ideals of B are just the graded primes $\{n_i : i \in I\}$ lying over m. The natural mapping $B \to \oplus Q^g_{n_i}(B)$ is a graded isomorphism (we write $\bar{B} = B \otimes R/m$).

Furthermore, the following conditions are equivalent:

(3.13.1) B is gr-decomposed.

(3.13.2) $B \to \oplus Q^g_{n_i}(B)$ is a graded isomorphism.

(3.13.3) The gr-decomposition of $\bar{B}$ can be lifted to one of B.

Proof: An easy adaptation of [53], Ch. I, Prop. 1, 2, 3. ■

II.3.14 Corollary A graded field is gr-henselian. ■

II.3.15 Proposition Let B be a finite graded algebra over the gr-local ring R; then the natural mapping $\pi : \mathrm{Idemp}(B) \to \mathrm{Idemp}(\bar{B})$ is injective. Moreover, B is gr-decomposed if π is bijective.

Proof: Let e, e' be idempotents of degree zero such that $\pi(e) = \pi(e')$, that is, such that $x = e - e' \in mB$; then $x^3 = (e - e')^3 = e^3 - 3e^2e' + 3ee'^2 - e'^3 = e - e' = x$, so $x(1 - x^2) = 0$. As $x \in mB \subset J^g(B)$, clearly $x^2 \in J^g(B)$, so $1 - x^2$ is invertible, implying that $x^2 = 0$, hence $x = 0$. For $1 \le i \le n$, let $\pi(e_i) = (0,\ldots,1,\ldots,0) \in B^\sim = \oplus Q^g_{n(i)}(B)$. Each idempotent of $\bar{B}$ is a sum of some of the $\pi(e_i)$ and this can be lifted to B_0 if and only if $Q^g_{n(i)}(B)$ is a direct factor of B. ■

Let us now give some other characterizations of gr-henselian rings. As it happens, R is gr-henselian if and only if R_0 is henselian.

II.3.16 Theorem Let R be a gr-local ring with gr-maximal ideal m and let $\bar{R} = R/m$; then the following conditions are equivalent:

(3.16.1) R is gr-henselian.

(3.16.2) Each graded free R-algebra is gr-decomposed.

(3.16.3) $R[X]/(P)$ is gr-decomposed, for any monic homogeneous polynomial $P \in R[X; \deg X = t]$.

(3.16.4) For each $t \in \mathbb{N}$ and for each monic homogeneous polynomial $P \in R[X; \deg X = t]$ which has the property that $\pi(P) = \pi(Q)\pi(S)$ for some monic homogeneous polynomials $\pi(Q), \pi(S) \in \bar{R}[X; \deg X = t]$ with $(\pi(Q), \pi(S)) = 1$, we have $P = QS$ for some monic homogeneous polynomials lifting $\pi(Q)$ and $\pi(S)$.

(3.16.6) For each monic polynomial $P \in R_0[X]$ with $\pi(P) = \pi(Q)\pi(S)$ for some monic polynomials $\pi(Q), \pi(S) \in R_0[X]$ with

$(\pi(Q), \pi(S)) = 1$, we have $P = QS$ for some monic polynomials lifting $\pi(Q)$ and $\pi(S)$.

(3.16.7) R_0 is henselian.

(3.16.8) Each monic polynomial $P \in R_0[X]$ such that the image $\pi(P) \in \bar{R}_0[X]$ has a simple root a in $\bar{R}_0$ has a root in R_0 lifting a.

(3.16.9) Each monic homogeneous polynomial $P \in R[X; \deg X = t]$ such that $\pi(P)$ has a simple root a in $h(\bar{R})$ has a root in $h(R)$ lifting a.

(3.16.10) If S is gr-étale extension of R and if $n \in \mathrm{Spec}^g(S)$ lies over m, and if $k^g(n) = \bar{R}$, then $R \cong_g Q_n^g(S)$.

Proof: The implications (1) $\Rightarrow$ (2) $\Rightarrow$ (3) $\Rightarrow$ (4), and (5) $\Rightarrow$ (6) are clear. The implications (3) $\Rightarrow$ (5), (4) $\Rightarrow$ (6), (6) $\Rightarrow$ (4), and (4) $\Rightarrow$ (1) are just easy modifications of the proof of the classical theorem given in [53], Ch. I, Proposition 5. The implication (6) $\Rightarrow$ (7) also follows from this theorem, and (7) $\Leftrightarrow$ (8) follows from [53], Ch. VII, Proposition 3. The implication (9) $\Rightarrow$ (8) is clear and (5) $\Rightarrow$ (9) follows from the observation that we have $\pi(P) = (X - a)\pi(Q)$ with $(X - a, \pi(Q)) = 1$.

(9) $\Rightarrow$ (10) We may suppose that S is gr-étale of standard type, that is, $S = (R[X]/(f))_g$, where f is a monic homogeneous polynomial. As $k^g(n) = \bar{R}$, it follows that n corresponds to a homogeneous root $\pi(a)$ of $\pi(f) \in k[X]$.

Because f' is invertible in S, $\pi(a)$ is a simple root of $\pi(f)$, so it may be lifted to a homogeneous root of f. Hence $f = (X - a)g$, where g is a monic homogeneous polynomial. We may therefore prove that $R[X]/(f) \cong_g R[X]/(X - a) \oplus R[X]/(g)$. Now, due to the choice of $\pi(a)$, the gr-maximal ideal n of $R[X]/(f)$ corresponds to $Q_n^g(S) \cong_g Q_n^g(R[X]/(f)) \cong_g R[X]/(x - a) \cong_g R$.

Finally, the implication (10) $\Rightarrow$ (1) may be derived as follows. Let S be a finite graded free R-algebra, with basis $\{e_1, \dots, e_r\}$. Take $b = \sum_{i \leq r} x_i e_i \in S_0$; then there exist homogeneous polynomials $P_1, \dots, P_r$ in $R[X_1, \dots, X_r; \deg X_i = -\deg e_i]$, such that b is an idempotent in S if and only if $P_i(x_1, \ldots, x_r) = 0$. Let J be the graded ideal generated by $P_1, \ldots, P_r$ and let $E = R[X_1, \ldots, X_r]/J$. It is then easy to see that there is a bijection between $\mathrm{Hom}_{\text{gr-}R\text{-alg}}(E, C)$ and the idempotents of $S \otimes_R C$, for any graded R-algebra C, given by sending some u to $\sum e_i \otimes u(x_i)$. Now, let e be an idempotent in $B \otimes_R \bar{R}$; then e corresponds to a homomorphism of degree zero $\bar{u} : E \to \bar{R}$. Let $n = \mathrm{Ker}(\bar{u})$; then we have a factorization of $\bar{u}$ as $E \to Q_n^g(E) \to \bar{R}$. We have to show that $\bar{u}$ may be lifted to $u : E \to R$, or equivalently, that we may lift the map $Q_n^g(E) \to \bar{R}$ to $Q_n^g(E) \to R$. By (10), $Q_n^g(E) \cong R$, so we can lift it indeed. ∎

II.3.17 Proposition If R is a gr-henselian ring and S is a graded R-algebra of finite type which is quasifinite at a graded prime ideal n lying over m, then $Q_n^g(S)$ is a finite R-algebra and a direct factor of S.

Proof: We know that there exists $f \in h(S-n)$ and a finite graded R-algebra S' such that $S'_f \cong S_f$. As R is gr-henselian, S' is a direct sum of gr-local rings, so $Q_n^g(S)$ is a direct factor of S' and of S'_f. So we have maps $S \to S_f \to Q_n^g(S)$ and $\mathrm{Spec}(Q_n^g(S)) \to \mathrm{Spec}(S)$ is an open immersion. Moreover, $Q_n^g(S)$ is finite over R and S, being a direct factor of S'. So $\mathrm{Spec}(Q_n^g(S)) \to \mathrm{Spec}(S)$ is closed and $Q_n^g(S)$ is a direct factor of S. ∎

An easy application of the foregoing results also yields Hensel's lemma for gr-local rings:

II.3.18 Proposition Every gr-complete gr-local ring is gr-henselian. ∎

Let us conclude this section by a brief study of strictly gr-henselian rings. We call a gr-local ring R *strictly gr-henselian* (or *strictly gr-local*) if it is gr-henselian and if R/m is gr-separably closed.

For a gr-local ring R, we call a *gr-henselization* of R any pair (R^{gh}, i), where R^{gh} is gr-henselian and $i : R \to R^{\mathrm{gh}}$ is a gr-local morphism such that for every gr-henselian ring S and for every gr-local morphism $u : R \to S$, there exists a unique gr-local morphism $u^{\mathrm{gh}} : R^{gh} \to S$ such that $u = u^{\mathrm{gh}} \circ i$.

Similarly, we call a *strict gr-henselization* of R any quadruple $(R^{\mathrm{sgh}}, \Omega, i, \alpha)$, where R^{sgh} is strictly gr-henselian, $i : R \to R^{\mathrm{sgh}}$ is a gr-local morphism, Ω a gr-separable closure of $\bar{R}$, and $\alpha : R^{\mathrm{sgh}} \to \Omega$ is a gr-local morphism such that for any strictly gr-henselian ring S which is a graded R-algebra with gr-local structural morphism $u : R \to S$; we have that for every graded field $\beta : S \to \Omega$ every $j : \Omega \to \Omega'$ with $j \circ \alpha \circ i = \beta \circ u$, there exists a unique gr-local morphism $u^{\mathrm{sgh}} : R^{\mathrm{sgh}} \to S$ such that the following diagram of gr-local R-algebras is commutative:

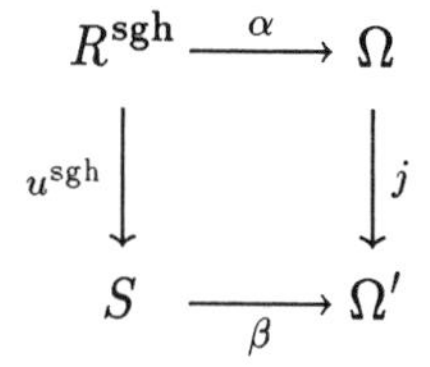

It is established immediately that R^{gh} and R^{sgh} (if they exist!) are unique up to graded isomorphism. We also have:

II.3.19 Proposition If (R_0^h, i_0) is a henselization of R_0, then $(R_0^h \otimes R, i_0 \otimes R)$ is a gr-henselization of R.

Proof: It is clear that $R_0^h \otimes R$ (tensor product over R_0) is gr-henselian because $(R_0^h \otimes R)_0 = R_0^h$ is henselian. Furthermore, if $u : R \to S$ is a gr-local morphism, where S is gr-henselian, then S_0 is henselian and $u_0 = u|_{R_0} : R_0 \to S_0$ is a local morphism, so there exists $u_0^h : R_0^h \to S_0$ such that $u_0 = u_0^h \circ i$. Define

$f : R_0^h \times R \to S$ by $f(a,b) = u_0^h(a)u(b)$. Clearly f is R_0-bilinear, so f corresponds to an R_0-linear map $u^{\mathrm{gh}} = R_0^h \otimes_{R_0} R \to S$, satisfying $u^{\mathrm{gh}}(a \otimes b) = u_0^h(a)u(b)$. But then $u = u^{\mathrm{gh}} \circ (i_0 \otimes R)$, and this finishes the proof. ■

II.3.20 Lemma Let R be a commutative ring; then R is strictly gr-henselian if and only if Spec(R) is connected and every gr-étale covering $R \to R'$ admits a section $R' \to R$.

Proof: Suppose R is strictly gr-henselian and let R' be a gr-étale covering of R; then R'/mR' is a gr-étale covering of $K = R/mR$, so is isomorphic to $K \oplus \cdots \oplus K$, since K is gr-separably closed. Now, take $f_i : R' \to R'/mR' \to K$ (ith projection) and let $n(i) = f_i^{-1}(0)$. By (3.16.10) it follows that $Q_{n(i)}^g(R') \cong R$, so $R \to R'$ admits a section.

Conversely, suppose that R is connected and that every gr-étale covering admits a section. Assume first that R is not gr-local; then there exists $\lambda, \mu \in h(R)$ such that λ and μ are not invertible, but $\lambda + \mu$ is. Let $R' = R_\lambda \times R_\mu$; then R' is a graded étale covering of R, so there exists a section $R' \to R$. Hence $R_\lambda \to R$ or $R_\mu \to R$, so λ or μ is invertible. Hence R is gr-local.

To show that R is gr-henselian we use (3.16.10). Suppose that R' is a gr-étale R-algebra, n a graded prime of R', and $R'/n \cong R/m$; then $R \to Q_n^g(R')$ is a gr-étale covering, so it has a section. By Lemma II.3.21 below, it then follows that $R \cong_g Q_n^g(R)$.

Finally, suppose that $K = R/m$ is not gr-separably closed. Let L be a gr-separable extension of K; then $L = K[a]$. Lift the (homogeneous) minimal polynomial f of a to a monic homogeneous polynomial f and let $R' = R[X]/(f)$; then clearly R' is gr-étale over R, so $R \to R'$ splits. Hence f has a root within R, so $\tilde{f}$ has a root within K. ■

II.3.21 Lemma If $R \to S$ is a gr-local morphism of gr-local rings which is gr-separable and admits a section $S \to R$, then $R \cong_g S$.

Proof: As S/R is graded separable, $S \otimes_R S \cong_g S \oplus S'$. Now, as R is a graded S-algebra, because of the section, $S \cong_g R \otimes_R S \cong_g R \otimes_S (S \otimes_R S) \cong_g R \otimes_S (S \oplus S') \cong_g R \times (R \otimes_S S')$. As S is gr-local, $S \cong_g R$ indeed. ■

II.3.22 Corollary Let $(R_i)_i$ be an inductive system of strictly gr-henselian rings, where the connecting morphisms are graded, but not necessarily gr-local; then the inductive limit $\varinjlim R_i$ is also strictly gr-henselian.

Proof: It suffices to verify that the conditions of (II.3.20) are invariant under inductive limits. ■

II.3.23 Theorem For every gr-local ring R, there exists a strict gr-henselization $(R^{sg}, \Omega, i, \alpha)$.

Proof: First suppose that R/m is nontrivially graded, that is, $R/m = k[T, T^{-1}]$, where we assume $\deg T = r$ if $\operatorname{char} k = 0$ and $\deg T = rp^n$, where $(r,p) = 1$ if $\operatorname{char} k = p$. Let $(R_0^{sh}, \Omega_0, i_0, \alpha_0)$ be a strict henselization of R_0 and put $R_1 = R_0^{sh} \otimes R$. Exactly as in (II.3.19) it follows that R_1 is gr-henselian and $R_1/mR_1 = \Omega_0[T, T^{-1}]$. Lift T to a homogeneous $t \in R_1$ and write $R_2 = R_1[X]/(X^r - t)$; then R_2 is a finite gr-étale extension of R_1 and has a connected spectrum (one may actually even prove that R_2/R_1 is gr-Galois, as R_1 has rth roots of unity!). So R_2 is gr-henselian. It is also easy to see that R_2 is strictly gr-henselian as $R_2/mR_2 = \Omega_0[X, X^{-1}]$ is gr-separably closed. The ring R_2 determines the strict gr-henselization of R, as soon as one has checked the universal property. This may be done as in (II.3.19); we leave details to the reader. ■

II.3.24 Note Following the techniques developed in [54], it is also possible to prove that R is gr-henselian if and only if the functor $F : S \to S \otimes_R K$ yields an equivalence between the category of finite gr-étale R-algebras (resp. gr-étale gr-local algebras) and the category of gr-étale K-algebras (resp. finite separable graded field extensions of K). It is also possible to describe R^{gh} and R^{sgh} as inductive limits of gr-local-étale R-algebras, where a gr-local-étale R-algebra is of the form $Q_n^g(S)$, where S/R is gr-étale and n lies over m.

II.4 THE JOIN OF GR-HENSELIAN RINGS

As the title of this section indicates, its contents are largely inspired by M. Artin's famous paper [2]. We were greatly helped by preliminary notes to a forthcoming publication on this topic by Ojanguren and Sridharan, which were kindly communicated to us by M. A. Knus.

Let us call a graded domain R *graded absolutely integrally closed* (gaic) if its graded field of fractions $K = Q^g(R)$ is gr-algebraically closed and if R is integrally closed in K. The *graded absolute integral closure* $R^\sharp$ of a graded domain R is the integral closure of R in the gr-algebraic closure of $Q^g(R)$. Finally, let us call a commutative graded ring graded absolutely integrally closed if it is the finite direct product of gaic domains.

Note that a domain R is gaic if and only if every monic homogeneous polynomial P may be written as a product $\prod_{i\leq n}(X - a_i)$, where $a_i \in h(R)$.

II.4.1 Lemma Let R be graded absolutely integrally closed; then

(4.1.1) $S^{-1}R$ is gaic, for every multiplicative subset of $h(R)$.

(4.1.2) R/p is gaic for every $p \in \mathrm{Spec}^g(R)$.

(4.1.3) If the domain S is a finite graded R-algebra, then $S \cong_g R/p$ for some $p \in \mathrm{Spec}^g(R)$.

Proof: The first two statements are obvious, using the foregoing remark and the fact that a localization of a normal domain is normal too. For the third statement, let $p = \mathrm{Ker}(R \to S)$, where $R \to S$ is the structural morphism; then p is a prime ideal of R, as S is a domain, so R/p is gaic and we have a monomorphism $R/p \to S$. Now S is integral over R/p and it has the same graded field of fractions, as R/p is gaic. It follows that this embedding is also surjective, hence an isomorphism. ■

II.4.2 Proposition Let R be a graded absolutely integrally closed domain. If $R \to S$ is gr-étale, then S is gaic, that is, $S = S_1 \times \cdots \times S_r$, where the $R \to S_i$ are graded open maps. Conversely, any such product of graded open extensions is gaic.

Proof: Recall that a map $\varphi : R \to S$ is open if and only if it induces an open immersion $\varphi^+ : \mathrm{Spec}(S) \to \mathrm{Spec}(R)$, or, equivalently, if φ is étale and if $S \otimes S \cong S$. Let K be the graded field of fractions of R; then $K \otimes S \cong_g K^r$. As S is normal, $S \cong_g \prod_{i \leq r} S_i$, for some étale $S \to S_i$. Finally, $S_i \otimes S_i \to S_i$ is an isomorphism of R and the S_i have the same graded field of fractions. Conversely, if $R \to S_i$ is open, then R and S_i have the same graded field of fractions and S_i is normal. ■

II.4.3 Corollary Let $\varphi : R \to S$ be a gr-étale morphism of graded domains with graded fields of fractions K and L. Suppose that an embedding $K^a \to L^a$ is given, where $(\_)^a$ denotes the gr-algebraic closure; then φ induces an open embedding $R \to S$.

Proof: The map $R^\sharp \to R^\sharp \otimes S$ is gr-étale, so $R^\sharp \otimes S = \prod_{i \leq r} S_i$, where $S \to S_i$ is open and S_i is a gaic domain. So we have a morphism $f : \prod_{i \leq r} S_i \to S^\sharp$, hence $S_i = S^\sharp$ for some i. Indeed,

$\mathrm{Ker}(f) = S_1 \times \cdots \times p_i \times \cdots \times S_r$, as $S^\sharp$ is a domain. So we have $R^\sharp \to \prod_{j \leq r} S_j / \mathrm{Ker}(f) = S_i/p_i \to S^\sharp$. Now $R^\sharp \to S^\sharp$ is open, so $\mathrm{Spec}(S_i) \to \mathrm{Spec}(R^\sharp)$ is injective, which implies that $p_i = 0$, so $R^\sharp \to S_i \to S^\sharp$. Now $S^\sharp$ is integral over R, over $R^\sharp \otimes S$, and over S_i. Let $K = Q^g(S_i)$; then $R^\sharp \to S \to S^\sharp$, and $S_i = S^\sharp$ as S_i is gaic. So $R \to S^\sharp = S_i$. ∎

II.4.4 Lemma Let R be a gr-semilocal ring and S a graded R-algebra which is projective of finite constant rank as an R-module; then S is graded free as an R-module and possesses a basis consisting of elements of degree zero.

Proof: The lemma is obvious when R is a finite product of graded fields. In the general case $\bar{R} = R/J^g(R)$ is a finite product of graded fields, so take a basis consisting of elements of degree zero in $\bar{S} = S \otimes \bar{R}$, and lift these to a homogeneous subset of S. An application of the graded version of Nakayama's lemma yields that this is a basis for S. ∎

II.4.5 Lemma Let R be a gr-semilocal ring and p and p' gr-maximal ideals, and suppose that $\mathrm{Spec}(R)$ is connected. Assume also that $R \times R$ is the only finite gr-étale extension R' of rank 2 of R splitting at p and p' (i.e., it is the only extension satisfying $R'/pR' = R/pR \times R/pR$ and $R'/p'R' = R/p'R \times R/p'R$). Then $X = \{P \in \mathrm{Spec}^g(R) : P \subset p \cap p'\}$ has a unique maximal element, that is, $Q^g_p(R) \otimes Q^g_{p'}(R)$ is a gr-local ring.

Proof: Let S be the homogeneous multiplicatively closed set generated by $h(R - (p \cap p'))$; then there is a natural (graded) isomorphism of the form $Q^g_p(R) \otimes Q^g_{p'}(R) \to S^{-1}R$ given by sending $(x/s) \otimes (x'/s')$ to xx'/ss'. Moreover, $\mathrm{Spec}^g(S^{-1}R) = \{P \in \mathrm{Spec}^g(R) : P \cap S = \Phi\} = \{P \in \mathrm{Spec}^g(R) : P \subset p \cap p'\}$. So, the fact that X has a unique maximal element implies that $Q^g_p(R) \otimes Q^g_{p'}(R)$ is gr-local indeed.

Observe also that $X = \Phi$ if and only if $S^{-1}R = 0$, or $Q_p^g(R) \otimes Q_{p'}^g(R) = 0$. Suppose this to be the case, indeed. Let $T = Q_p^g(R) \times Q_p^g(R)$; then T is clearly faithfully flat and $T \otimes T = (Q_p^g(R) \otimes Q_{p'}^g(R)) \otimes (Q_p^g(R) \times Q_{p'}^g(R)) = T = R \otimes T$, so $R \cong T$ and T is not connected. Hence we may assume that $X \neq \Phi$.

By Zorn's lemma, X has a maximal element. Suppose that P_1 and P_2 are two maximal elements of X. We construct a gr-étale extension of R of rank 2 splitting at p and p'. Define $R_* = R/P_1 \cap P_2$; then we have a fibred product

$$\begin{array}{ccc} R_* & \longrightarrow & R/P_2 \\ \downarrow & & \downarrow \\ R/P_1 & \longrightarrow & R/P_1 + P_2 \end{array} .$$

Consider the gr-étale extension $R/P_i \times R/P_i$ of R/P_i. As there is no $q \in \operatorname{Spec}(R/P_1 + P_2)$ with $q \subset p \cap p'$, we have that $Q_p^g(R/P_1 + P_2) \otimes Q_{p'}^g(R/P_1 + P_2) = 0$ and $R/P_1 + P_2 \cong_g Q_p^g(R/P_1 + P_2) \times Q_{p'}^g(R/P_1 + P_2)$ by an argument similar to the one above. Now, define R'_* as the pullback in the following diagram (R'_* is graded, as the connecting maps have degree zero):

$$\begin{array}{ccc} R'_* & \longrightarrow & R/P_2 \times R/P_2 \\ \downarrow & & \downarrow g \\ R/P_1 \times R/P_1 & \xrightarrow{f} & (R/P_1 + P_2) \times (R/P_1 + P_2) \end{array} .$$

Observe that $(R/P_1 + P_2) \times (R/P_1 + P_2) \cong_g Q_p^g(R/P_1 + P_2) \times Q_{p'}^g(R/P_1 + P_2) \times Q_p^g(R/P_1 + P_2) \times Q_{p'}^2(R/P_1 + P_2)$, and define $f(a,b) = (a,a,b,b)$ and $g(x,y) = (x,y,y,x)$. By gr-localizing the foregoing diagrams at p and p', we see that R'_* is a graded extension of rank 2 of R_*. It is also clear that $\operatorname{Spec}(R'_*)$ is connected, since the idempotents of R'_* are just the (e_1, e_2) with e_i

idempotent in $R/P_i \times R/P_i$ and such that $f(e_1) = g(e_2)$, that is, $\mathrm{Idemp}(R'_*) = \{(0,0),(1,1)\}$. Since it is easily seen that R'_* is a gr-étale extension of R_* and that R'_* is graded free by II.4.4, we have $R'_* = R_* \oplus R_*x$, hence $R'_* = R_*[X]/(X^2 + b_0X + c_0)$ for some $b_0, c_0 \in (R_*)_0$. Lift b_0, c_0 to $b, c \in R_0$ and define $R' = R[X]/(X^2 + bX + c)$; then R' has the following properties:
(i) R' splits at p and p', as $R'/pR' = R'_*/pR'_*$.
(ii) R' is connected, since a nontrivial idempotent of R' induces a nontrivial idempotent of R'_*, because $P_1 \cap P_2 \subset p \cap p' \subset J^g(R')$.
(iii) R' is a gr-étale R-algebra.
Let $f = X^2 + bX + c$; then we show that f' is invertible in R, or, equivalently, $(f')^2 = (2X + b)^2$ is invertible. Now, $(2X + b)^2 = 4(X^2 + bX + c) + b^2 - 4c = (b^2 - 4c)\mathrm{mod}(X^2 + bX + c)$. As R'_* is a gr-étale extension of R_*, it follows that $b_0^2 - 4c_0$ is invertible, and so is $b^2 - 4c$, as $P_1 + P_2 \subset J^g(R)$. ■

II.4.6 Corollary If R is a gaic domain and $p, q \in \mathrm{Spec}^g(R)$, then the subring of K generated by $Q_p^g(R)$ and $Q_q^g(R)$, which we will denote $[Q_p^g(R), Q_q^g(R)]$, is gr-local.

Proof: Let S be the multiplicative set $h(R - (p \cup q))$, and replace R by $S^{-1}R$; then R has two gr-maximal ideals p, q. If $[Q_p^g(R), Q_q^g(R)] = Q_p^g(R) \otimes Q_q^g(R)$ is not gr-local, then there exists a connected gr-étale extension R' of R splitting at p and q. By II.4.2, $R' \cong_g \prod_{i \leq r} R_i$, where $R \to R_i$ is open and R_i is gaic. But $\mathrm{rk}(R') = 2$, so $r = 2$ and $R' = R \times R$. ■

II.4.7 Corollary If R is a gaic ring and $p, q \in \mathrm{Spec}^g(R)$, then $p + q = R$ or $p + q$ is prime.

Proof: We may assume R to be a domain. Suppose $p + q \neq R$; then we first show that $\mathrm{rad}^g(p+q) = \cap\{x \in \mathrm{Spec}^g(R) : p + q \subset x\}$ is prime. Choose two maximal graded primes p' and q', maximal with respect to the properties $p \subset p' \subset p_1 \cap p_2$ and $q \subset q' \subset p_1 \cap p_2$.

Then clearly p' and q' do not contain $p+q$, so $p' \neq q'$. Now gr-semilocalize at p_1 and p_2; then p_1 and p_2 become gr-maximal ideals. But then p' and q' are two different graded prime ideals, maximal among those contained in $p_1 \cap p_2$. By II.4.5, there exists a graded étale extension of rank two splitting at p' and q'. An argument similar to that which we used in II.4.6 shows this to be impossible. Next, let us show that $p+q$ is prime. Indeed, suppose $a^2 \in h(p+q)$; then $a^2 = u_1 + u_2$, with $u_1 \in h(p)$ and $u_2 \in h(q)$. Choose $x \in h(R)$ such that $x^2 = u_1$ (R is gaic!); then $x \in p_1$ and $a^2 = x^2 + u_2$ and $u_2 = (a+x)(a-x) \in q$, so $a = x + (a-x) \in p+q$, or $a = -x + (a+x) \in p+q$. ■

II.4.8 Proposition A graded normal domain is gr-henselian if and only if $R^\sharp$ is gr-local.

Proof: Just mimic the proof of the analogous ungraded statement. ■

From now on, let us assume R to be a normal graded domain and let p and q be graded prime ideals of R; we denote by K^a the graded algebraic closure of $K = Q^g(R)$, and R_p^{gh} and R_q^{gh} are the gr-henselizations of the graded localizations of R at p and q. We fix two embeddings, $i_p : R_p^{\mathrm{gh}} \to K$ and $i_q : R_q^{\mathrm{gh}} \to K$.

II.4.9 Theorem With notations as above, $S = [(R_p^{\mathrm{gh}}, R_q^{\mathrm{gh}})]$ is gr-henselian.

Proof: The embeddings $R \to R_p^{\mathrm{gh}} \to S \to K^a$ and $R \to R_q^{\mathrm{gh}} \to S \to K^a$ induce, by taking the graded absolute integral closure, embeddings $R^\sharp \to (R_p^{\mathrm{gh}})^\sharp \to S^\sharp \to K^a$, resp. $R^\sharp \to (R_q^{\mathrm{gh}})^\sharp \to S^\sharp \to K^a$, and we know that $(R_p^{\mathrm{gh}})^\sharp$ is a union of open extensions of $R^\sharp$, so it is gr-local. Hence, $(R_p^{\mathrm{gh}})^\sharp = Q_P^g(R^\sharp)$, for some graded prime ideal P of $R^\sharp$ lying over p. It is also clear that $(R_q^{\mathrm{gh}})^\sharp$ is gr-local, so $[(R_p^{\mathrm{gh}})^\sharp, (R_q^{\mathrm{gh}})^\sharp]$ is gr-local. Now, $S \subset [(R_p^{\mathrm{gh}})^\sharp, (R_q^{\mathrm{gh}})^\sharp] \subset S^\sharp$ and

$[(R_p^{\mathrm{gh}})^\sharp, (R_q^{\mathrm{gh}})^\sharp]$ is gaic, since it is a gr-henselization of $R^\sharp$, so $S^\sharp$ is gr-local and S is gr-henselian. ■

II.4.10 Lemma If in II.4.9 the graded prime ideals p and q do not contain each other, then $A = R_p^{\mathrm{gh}} \cap R_q^{\mathrm{gh}}$ is an integrally closed gr-semilocal domain containing two maximal ideals pA and qA. Moreover, A is a direct limit of gr-étale R-algebras, and A is noetherian if R is noetherian.

Proof: First, choose $x_1, \ldots, x_n \in h(A)$ and let R' be the integral closure of R in the graded field K' generated over K by $x_1, \ldots, x_n$. As R_p^{gh} is normal, $R' \subset A$, so A is normal. Next, let $p' = pR_p^{\mathrm{gh}} \cap R'$; then we may show that $R'^{\mathrm{gh}}_{p'} = R_p^{\mathrm{gh}}$, using the fact that $Q_{p'}^g(R') \subset R_p^{\mathrm{gh}}$.

Now, consider $Q_{p',q'}^g(R')$, the gr-semilocalization of R at p' and q', which is the localization of R' at the multiplicatively closed subset $h(R' - (p' \cup q'))$; then $Q_{p',q'}^g(R')$ is gr-étale over R at p' and q'. Now $A = \cup Q_{p',q'}^g(R')$, as one easily verifies. Indeed, let $x \in h(A)$ and let R', K', be as above; then $K' = Q^g(R')$ and $x \in K' \cap Q_{p',q'}^g(R') = K' \cap R_p^{\mathrm{gh}}$. We have $K' \cap R'^{\mathrm{gh}}_{p'} = Q_{p'}^g(R')$ and $x \in Q_{p'}^g(R')$. Similarly, $x \in Q_{q'}^g(R')$, so $x \in Q_{p',q'}^g(R') = Q_{p'}^g(R') \cap Q_{q'}^g(R')$.

We are left to show that A is noetherian if R is noetherian. This is easily seen, using the fact that an inductive limit of gr-local noetherian rings is again noetherian, if the connecting homomorphisms are flat and gr-local, and for each pair of indices i, j we have $m_i R_j = m_j$. ■

II.4.11 Lemma With notation as before and $pA = p$, $qA = q$, $[A_p^g, A_q^g]$ is gr-local.

Proof: If not, then A has a nontrivial gr-étale extension A' of rank two splitting at p and q. We also know that A' is a domain, as A is normal, so $A' \subset K$. Since A' splits over p, we obtain

$R_p^{\text{gh}} \otimes_A A' \cong_g R_p^{\text{gh}} \times R_p^{\text{gh}}$, so $A' \subset R_p^{\text{gh}}$, because otherwise $A' = A[x]$, where $x \notin R_p^{\text{gh}}$ and $R_p^{\text{gh}} \otimes_A A' \cong_g R_p^{\text{gh}}[x]$. So $A' \subset R_p^{\text{gh}} \cap R_q^{\text{gh}}$ and $A' \subset A$, a contradiction. ■

II.4.12 Theorem With notations as above, assume that the graded prime ideals p and q of R do not contain each other; then the graded residue class field $\bar{S}$ of $S = [R_p^{\text{gh}}, R_q^{\text{gh}}]$ is of the form k or $k[T,T^{-1}]$, where k is separably closed.

Proof: (4.12.1) It is sufficient to show that the separable closure of k is finite over k. Indeed, in this case, k is separably closed unless it is a real closed field. The latter is impossible: if the characteristic of the residue class field of A_p^{gh} is $l > 0$, and $\text{char}(S) = 0$, then the l-adic integers $\mathbf{Z}_l$ are contained in A_p^{gh}, so -1 is a sum of squares in A_p^{gh}, S, and k.

If the graded residue class fields of A_p^{gh} and A_q^{gh} have zero characteristic, then, by the Chinese remainder theorem, we may choose $a \in h(R)$ such that $h = 1 \operatorname{mod} p$ and $a = -1 \operatorname{mod} q$. By II.3.16, a and $-a$ are units and have square roots in A_p^{gh} and A_q^{gh} respectively. So -1 has a square root in S and k.

(4.12.2) Next, by II.3.22, we may assume that R is of finite type over $\mathbf{Z}$, that is, $R = \mathbf{Z}[x_1, \ldots, x_n]$, where the x_i are homogeneous. Indeed, if R is integrally closed, then it is a union of some A_i, which are graded integrally closed and of finite type over $\mathbf{Z}$, and the integral closure of a $\mathbf{Z}$-algebra of finite type is again of finite type.

(4.12.3) The proof may be reduced to the case where S is a graded field, that is, $S = k$ or $S = k[T,T^{-1}]$. Let B, $B^{(p)}$, resp. $B^{(q)}$, denote the integral closures of $R/m \cap R$, $R_p^{\text{gh}}/m \cap R_p^{\text{gh}}$, resp. $R_q^{\text{gh}}/m \cap R_q^{\text{gh}}$, in their graded fields of fractions, m denoting the gr-maximal ideal of S; then $B^{(p)}$ is the gr-henselization of B at a graded prime ideal P. To show this, observe first that $B^{(p)}$ is gr-henselian, because $R_p^{\text{gh}}/m \cap R_p^{\text{gh}}$ is gr-henselian, being a quotient of R_p^{gh}, and $B^{(p)}$ is a direct limit of finite graded algebras over $R_p^{\mathrm{gh}}/m \cap R_p^{\mathrm{gh}}$, with gr-local morphisms.

Let p' be the gr-minimal ideal of $B^{(p)}$ and let $P = p' \cap B$; then there are inclusions $B \to B_P^{\mathrm{gh}} \to B^{(p)} \to \bar{S}$. Since m is the gr-maximal ideal of $S = [R_p^{\mathrm{gh}}, R_q^{\mathrm{gh}}]$, we have $m \cap R_p^{\mathrm{gh}} \subset pR_p^{\mathrm{gh}}$ and $m \cap R_q^{\mathrm{gh}} \subset qR_q^{\mathrm{gh}}$, so $m \cap R \subset p \cap q$. Consider $R_p^{\mathrm{gh}}/(m \cap R)R_p^{\mathrm{gh}}$; then there is a commutative diagram

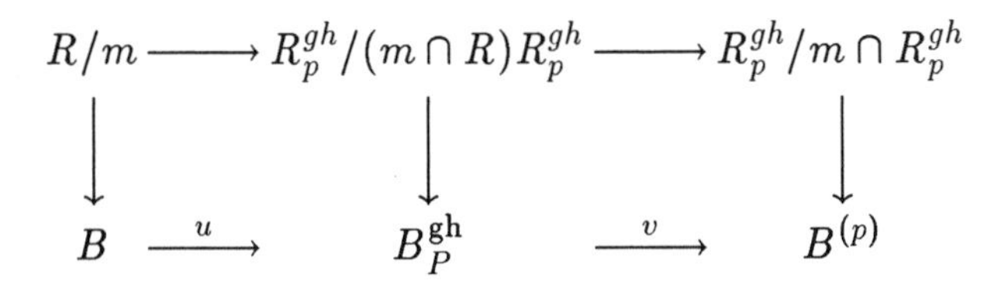

It follows that $\mathrm{Im}(v)$ contains $R_p^{\mathrm{gh}}/(m \cap R)R_p^{\mathrm{gh}}$. But B_P^{gh} is normal, so $\mathrm{Im}(v)$ contains the normalization of $R_p^{\mathrm{gh}}/(m \cap R)R_p^{\mathrm{gh}}$, hence $B_P^{\mathrm{gh}} = B^{(p)}$. It follows that $\bar{S} = [R_p^{\mathrm{gh}}/(m \cap R)R_p^{\mathrm{gh}}, R_q^{\mathrm{gh}}/(m \cap R)R_q^{\mathrm{gh}}] = [B^{(p)}, B^{(q)}], = [B_P^{\mathrm{gh}}, B_Q^{\mathrm{gh}}]$, which finishes our reduction.
(4.12.4) We suppose that S is a graded field and continue our proof, using induction on the graded Krull dimension gr K $\dim(C)$ (cf. [48]), where $C = [A_p^g, A_q^g]$. By II.4.10 and II.4.11, C is noetherian and gr-local. First, suppose gr K $\dim(C) = 0$, that is, C is a graded field. Take $a \in h(T)$, where T is the gr-separable closure of S; then a is algebraic over K, so there exists $\lambda \in R$ such that λa is integral over R and over A. Since A is integrally closed, the minimal polynomial f for a over $Q^g(A) = L$ is a monic polynomial in $h(A[X, \deg X = \deg a = \alpha])$. Using an argument due to Schmidt, we show that f may be decomposed completely in S.

Let $a_1, \ldots, a_n$ be the n distinct homogeneous roots of f in T and let $\delta = \pm N_{L[a]/L}(f'(a)) = \prod_{i<j}(a_i - a_j)^2$ be the discriminant of f. We claim that $\delta = \delta_p \delta_q$, where $\delta_p \in h(A - q)$ and $\delta_q \in h(A - p)$. If $\delta \in p \cap q$, this is trivial: let $\delta_p = \delta$ and $\delta_q = 1$ or $\delta_p = 1$ and $\delta_q = \delta$. So suppose $\delta \notin p \cap q$ and consider the primary decomposition $(\delta) = q_1 \cap \cdots \cap q_r$. Note that the q_i are graded ideals,

as δ is homogeneous. The graded prime ideals p_i associated to q_i lie within p or within q, but not within $p \cap q$, because $C = [A_p^g, A_q^g]$ is a graded field and $p \cap q$ contains no graded prime ideals different from zero. So $(\delta) = (q_1 \cap \cdots \cap q_s) \cap (q_{s+1} \cap \cdots \cap q_r) = a \cap b$, where $q_i \subset p$ for $i \leq s$ and $q_j \subset q$ for $j > s$. If a and b are principal ideals, then $\delta = ab$, and we are done. It is sufficient to check this gr-locally and, for example, $Q_p^g(a)$ is a principal ideal in $Q_p^g(A)$, since $Q_p^g(\delta) = Q_p^g(a) \cap Q_p^g(b) = Q_p^g(a)$.

Next, choose $g \in h(A[X, \deg(X) = \alpha])$ of degree n, which is fully decomposable in A, having n distinct roots of degree $\alpha : g(X) = \prod_{i \leq n}(X - \alpha_i)$; then the discriminant ξ of g is given by $\xi = \prod_{i<j}(\alpha_i - \alpha_j)^2 = \xi_p \xi_q$, by the argument used above. Consider the ideals $p\delta_p^2$ and $q\xi_q^2$ of A; then $p\delta_p^2 + q\xi_q^2 = 1$, because $\delta_p \notin q$, $\xi_q \notin p$, and p and q do not contain each other. By the Chinese remainder theorem, we may therefore find a homogeneous monic polynomial Θ in $A[X, \deg(X) = \alpha]$ such that $\Theta = g \bmod q\xi_q^2$ and $\Theta = f \bmod p\delta_p^2$. Clearly $\Theta \in A[X] \subset R_q^{\mathrm{gh}}[X]$ and the roots of g in A_q^{gh} are different modulo $q\xi$, ξ being the discriminant of g. Therefore $\Theta \bmod q\xi_q^2$ is fully decomposable into distinct linear factors in $R_q^{\mathrm{gh}}[X]$. By the graded version of Hensel's lemma, Θ splits into linear homogeneous factors in $R_q^{\mathrm{gh}}[X]$. Adjoin the roots of Θ to R_p^{gh} and take the normalization of this graded ring; then we obtain a gr-henselian ring D, with gr-maximal ideal m, say, and $D \subset S$. Now in $D[X]$ we have $\Theta = f \bmod \delta^2 m$. As Θ splits in $D[X]$, so does f. Use the fact that the roots are distinct modulo $\delta^2 m$, as δ is the discriminant of f. This finishes the proof for the case gr K $\dim(C) = 0$.

(4.12.5) Assume gr K $\dim(C) = r > 0$, and take $x \neq 0$ in $h(m)$, where m is the gr-maximal ideal of C. As S is a graded field, $x^{-1} \in S = [R_p^{\mathrm{gh}}, R_q^{\mathrm{gh}}]$, so $x^{-1} = \sum_{i \leq k} u_i v_i$, where $u_i \in h(R_p^{\mathrm{gh}})$ and $v_i \in h(R_q^{\mathrm{gh}})$. Let K' be the graded field generated by K, u_i, and v_i, where $K = Q^g(R)$, and let R' be the integral closure of R in K'. Put $p' = pR_p^{\mathrm{gh}} \cap R'$ and $q' = qR_q^{\mathrm{gh}} \cap R'$. Define A', S', C' in a similar way as A, S, C; then $u_i \in R_p^{\text{gh}} \cap K' \subset R'^g_{p'} \subset A'^g_{p'} \subset C'$, and $v_i \in C'$ in a similar way, so $x^{-1} \in C'$. Now, if the gr-separable closure of S' is a finite extension of S', then the same property holds for S. Indeed, $R'^{\text{gh}}_{p'}$ is a finite extension of R_p^{gh}, which may be seen as follows: $R' \otimes_R R_p^{\text{gh}}$ is finite over R_p^{gh}, so it is a product of gr-local rings, of which $R'^{\text{gh}}_{p'}$ is one of the factors. It follows that S' is finite over S.

Finally, gr K $\dim(C')$ < gr K $\dim(C)$. Indeed, first observe that $C' = [A'^g_{p'}, A'^g_{q'}]$ is a direct limit of gr-étale R'-algebras, by II.4.10. Let us show that it is also a limit of gr-étale $R' \otimes S[x^{-1}]$-algebras.

Let $C' = \lim C_i$, where C_i/R' is gr-étale; then we have a map $C_i \otimes_R C[x^{-1}] \to C'$. Clearly $C_i \otimes_R C[x^{-1}]$ is normal. There exists a direct factor $D_i[x^{-1}]$ of $C_i \otimes C[x^{-1}]$, so we have the following sequence of maps:

$$R' \otimes_R C[x^{-1}] \longrightarrow C_i \otimes_R C[x^{-1}] \longrightarrow D_i[x^{-1}] \longrightarrow C'.$$

Clearly, $C' = \lim D_i[x^{-1}]$ and gr K $\dim(D_i[x^{-1}]) \leq$ gr K dim $(R' \otimes_R C[x^{-1}])$; hence gr K $\dim(C') \leq$ gr K $\dim(R' \otimes_R C[x^{-1}]) =$ gr K $\dim(R' \otimes_R C[x^{-1}]) \leq$ gr K $\dim(C) - 1 <$ gr K $\dim(C)$. The equality holds, because $R' \otimes_R C[x^{-1}]$ is an integral extension of $C[x^{-1}]$. This finishes the proof. ∎

We call a graded commutative ring R *gr-acyclic* if every gr-étale covering $R \to S$ admits a section $S \to R$.

II.4.13 Proposition R is gr-acyclic if and only if every connected component of R is strictly gr-henselian.

Proof: This follows immediately from II.3.20. Recall that for $p \in \operatorname{Spec}(R)$, the connected component of R containing p is defined to be $\varinjlim R/eR$, where e runs through the inductive set of idempotents contained in p. Clearly every component is connected and if R is a graded ring, then so is every one of its components. ∎

A commutative graded ring, every connected component of which is gr-henselian, will be called *gr-quasiacyclic.* One easily checks that a direct limit of gr-(quasi)acyclic rings is again gr-(quasi)acyclic. It is also clear that a finite algebra over a gr-(quasi)acyclic ring is gr-(quasi)acyclic.

Before we come back to the main result of this section, let us briefly recall a definition from [10]. A commutative graded ring R is called *quasistrongly graded* if there exists a graded étale covering S of R containing a unit of degree one. In Chapter V, it will turn out that this is equivalent to the notion of being *strongly graded on the graded étale site.* Let us point out that these rings occur rather frequently.

II.4.14 Proposition The commutative ring R is quasistrongly graded in each of the following cases:
(4.14.1) R contains an invertible element of degree d such that d is invertible in R;
(4.14.2) R is strongly graded;
(4.14.3) R contains a generalized Rees ring, a scaled Rees ring, or a lepidopterous ring (cf. [57, 65, 66]), where $\deg X = d$, with d invertible in R.

Proof: In the first case, take $S = R[X]/(X^d - 1)$. In the second, let S_0 be an étale covering of R_0 such that $S_0 \otimes I \cong S_0$ (tensor product over R_0); then $S = S_0 \otimes R \cong_g S_0[X, X^{-1}]$ contains a unit of degree one. Finally, the third example reduces to an easy combination of the arguments used in the first cases. ■

II.4.15 Theorem Let R be a commutative ring, and let $p_1, \ldots, p_r \in \mathrm{Spec}^g(R)$.
(4.15.1) For $i = 1, \ldots, r$ let S_i be a direct limit of gr-étale extensions of $R^{\mathrm{gh}}_{p_i}$ such that $S_i \subset R^{\mathrm{sgh}}_{p_i}$. Then $S_1 \otimes_R \cdots \otimes_R S_r$ is gr-quasiacyclic; in particular, $R^{\mathrm{gh}}_{p_1} \otimes_R \cdots \otimes_R R^{\mathrm{gh}}_{p_r}$ is gr-quasiacyclic.

(4.15.2) If in (4.15.1) we have $r = 2$, $p_1 = p$, $p_2 = q$, and p and q do not contain each other, then the graded residue fields of all the components of $R_p^{\mathrm{gh}} \otimes_R R_q^{\mathrm{gh}}$ have a separably closed part of degree zero.
(4.15.3) The same property is also valid for the components of $R_{p_1}^{\mathrm{sgh}} \otimes_R \cdots \otimes_R R_{p_r}^{\mathrm{sgh}}$.
(4.15.4) If R is quasistrongly graded, then $R_{p_1}^{\mathrm{sgh}} \otimes_R \cdots \otimes_R R_{p_r}^{\mathrm{sgh}}$ is gr-acyclic.

Proof: It is easily seen, using induction on r, that we may restrict attention to $r = 2$, that is, we work with two graded prime ideals p and q. Write $R = \varinjlim R_i$, where R_i is a graded **Z**-algebra of finite type. If we write $p(i) = p \cap R_i$ and $q(i) = q \cap R_i$, then $R_p^{\mathrm{gh}} = \varinjlim (R_i)_{p(i)}^{\mathrm{gh}}$. If p and q do not contain each other, then the same property holds for $p(i)$ and $q(i)$, for any index i which is large enough, so, by the remarks preceding the theorem, we may restrict attention to the case where R is of finite type over **Z**, that is, R is a quotient of polynomial ring R_1 over **Z** and $T = R_p^{\mathrm{gh}} \otimes_R R_q^{\mathrm{gh}}$ is a quotient of $T_1 = (R_1)_{p(1)}^{\mathrm{gh}} \otimes_{R_1} (R_1)_{q(1)}^{\mathrm{gh}}$. Hence, by the remarks above, we may replace R by R_1, so we may assume R to be normal. As T is a limit of gr-étale algebras, each component B of T is a limit of gr-étale R-algebras, hence is a normal domain. It follows that B may be embedded into the graded algebraic closure K^a of $K = Q^g(R)$. The maps $R_p^{\mathrm{gh}} \to B$ and $R_q^{\mathrm{gh}} \to B$ generate B within K^a, so $B = [R_p^{\mathrm{gh}}, R_q^{\mathrm{gh}}]$. The statements (4.15.1) and (4.15.2) thus easily follow from II.4.9 and II.4.12.

For the proof of (4.15.3), let us write R^s for $R_p^{\mathrm{sgh}} \otimes_R R_q^{\mathrm{sgh}}$, and let B^s be a component of T^s. Embed T^s in K^a, where K^a is as above; then there are maps $R \to R_t^{\mathrm{sgh}} \to B^s \to K^a$, $t \in \{p, q\}$. By (4.15.1), B^s is gr-henselian, so we are left to show that the graded residue field of B^s has a separably closed part of degree zero. If $R_p^{\mathrm{sgh}} \subset R_q^{\mathrm{sgh}}$, then $B^s = R_q^{\mathrm{sgh}}$, and we are done. So we may assume that R_p^{sgh} and R_q^{sgh} do not contain each other.

Let $C(p)$ be the integral closure of R in R_p^{sgh}; then R_p^{sgh} is just the graded localization $Q_P^g(C(p))$ for some $P \in \mathrm{Spec}^g(C(p))$. Let $D = [C(p), C(q)]$; then D is integral over $C(p)$, so there is a unique graded prime ideal P'' in D lying over P and, similarly, a unique Q'' lying over Q. Assume $Q'' \subset P''$; then $D_{P''}^g \subset D_{Q''}^g$. Now, R_q^{sgh} is the union of all subrings of $D_{Q''}^g$, which are gr-étale over A, and similarly for R_p^{sgh}. But then $R_p^{\mathrm{sgh}} \subset R_q^{\mathrm{sgh}}$; hence we may assume that P'' and Q'' do not contain each other. There exists a finite normal graded extension R' such that $R \subset R' \subset D$, and $P' = P'' \cap R'$ and $Q' = Q'' \cap R'$ do not contain each other. As in the proof of II.4.12, we may replace R by R'. Now $B^s = [R_p^{\mathrm{sgh}}, R_q^{\mathrm{sgh}}]$ is integral over $B = [R_p^{\mathrm{gh}}, R_q^{\mathrm{gh}}]$, which has the desired property by II.4.12. So B^s has also a graded residue field with separably closed part of degree zero.

To prove (4.14.4), let $k[X, X^{-1}]$, $l[Y, Y^{-1}]$, and $F[T, T^{-1}]$ be the graded residue fields of R_p^{sgh}, R_q^{sgh}, and B^s. We already know that k, l, and F are separably closed. As R is quasistrongly graded, $\deg X = \deg Y = 1$, so $\deg T = 1$ and B^s is strictly gr-henselian. ∎

II.4.16 Example The conclusion of (4.15.4) does not hold for arbitrary graded rings, as the following example shows. Let R_0 be the ring of algebraic integers in a number field and p_1 a prime number such that the ideal $p_1 R_0$ is a product of two prime ideals P and Q. Let $R = R_0[T, T^{-1}]$, where $\deg T = p_1$, $p = PR$, and $q = QR$. The graded residue class fields of R_p^g and R_q^g have characteristic p_1, so they are of the form $k[T, T^{-1}]$, where $\deg T = p_1$, and similarly for R_p^{sgh} and R_q^{sgh}, so the join $[R_p^{\mathrm{sgh}}, R_q^{\mathrm{sgh}}]$ contains only units of degree $rp_1, r \in \mathbf{Z}$. Clearly $[R_p^g, R_q^g] \subset [R_p^{\mathrm{sgh}}, R_q^{\mathrm{sgh}}]$, and $[R_p^g, R_q^g]$ is the localization of R at the homogeneous multiplicatively closed set generated by $h((R - p) \cup (R - q))$, by an earlier remark. The prime number p_1 lies in this set, so $1/p_1 \in [R_p^g, R_q^g] \subset B^s$, and

B^s contains the rationals. So the graded residue field of B^s has characteristic zero, and it is not gr-separably closed, as it does not contain a unit of degree one!

II.4.17 Theorem (Graded version of Artin's Refinement Theorem) Let R be a noetherian quasistrongly graded ring, let $p(1),\ldots,p(r) \in \mathrm{Spec}^g(R)$, and let $R \to S$ be a gr-étale map. Write $R(p)$ for $R^{\mathrm{sgh}}_{p(1)} \otimes_R \cdots \otimes_R R^{\mathrm{sgh}}_{p(r)}$ and assume that T is a gr-étale covering of $S^{(n)} \otimes_R R(p)$; then there exists a gr-étale covering S' of S such that we have a factorization $S^{(n)} \otimes_R R(p) \to T \to S'^{(n)} \otimes_R R(p)$.

Proof: We use induction on n. For $n = 0$, the result follows from (4.15.4), so suppose the theorem is true for $0, 1, \ldots, n-1$. Choose a gr-étale map $S \to S_1$ such that we have a factorization $S^{(n)} \otimes_R R(p) \to T \to S_1^{(n)} \otimes_R R(p)$ (such an S_1 exists!). Let G be the image of the induced map $\mathrm{Spec}^g(S_1) \to \mathrm{Spec}^g(S)$ and suppose $G \neq \mathrm{Spec}^g(S)$. Take $q \in \mathrm{Spec}^g(S) - G$ and consider the inverse image p of q in $\mathrm{Spec}^g(R)$; then $S_q^{\mathrm{sgh}} = R_p^{\mathrm{sgh}}$, as S/R is gr-étale. Consider $S_1 \times R_p^{\mathrm{sgh}}$. The nth tensor power $(S_1 \times R_p^{\mathrm{sgh}})^{(n)}$ is a direct product of factors of the form $S_1^{(i)} \otimes_R (R_p^{\mathrm{sgh}})^{(j)}$, where $i + j = n$. First let $i < n$ and let T' be the pushout in the following diagram:

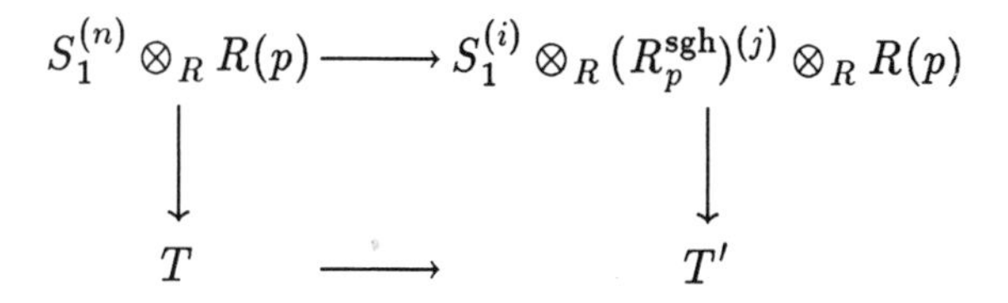

Clearly T' is graded, as the connecting maps are of degree zero. By the induction hypothesis, there exists a gr-étale covering S_2 of S and S_1, and a factorization

$$S^{(i)} \otimes_R (R_p^{\mathrm{sgh}})^{(j)} \otimes_R R(p) \longrightarrow T' \longrightarrow S_2^{(i)} \otimes_R (R_p^{\mathrm{sgh}})^{(j)} \otimes_R R(p),$$

so

$$S^{(n)} \otimes_R R(p) \longrightarrow T \longrightarrow S_2^{(i)} \otimes_R (R_p^{\text{sgh}})^{(j)} \otimes_R R(p).$$

We can choose S_2 such that this works for all $i < n$ at once. For $i = n$, it is also true, by (4.15.4.), so we have a factorization

$$S^{(n)} \otimes_R R(p) \longrightarrow T \longrightarrow (S_2 \otimes R_p^{\text{sgh}})^{(n)} \otimes_R R(p).$$

As R_p^{sgh} is a limit of gr-étale extensions of R, there exists a gr-étale extension S_2' of R such that

$$S^{(n)} \otimes_R R(p) \longrightarrow T \longrightarrow (S_2 \otimes S_2')^{(n)} \otimes_R R(p).$$

It is clear that $R \to S_2 \to S_2'$ is gr-étale and the image G_2 of $\text{Spec}^g(S_2 \times S_2')$ in $\text{Spec}^g(R)$ strictly contains G, as it contains q. If we repeat this, we obtain an increasing sequence of subsets $G = G_1 \subsetneq G_2 \subsetneq G_3 \subsetneq \cdots \subsetneq \text{Spec}^g(R)$. Since $\text{Spec}^g(R)$ is a noetherian topological space, there exists a positive integer n such that $G_n = \text{Spec}^g(R)$. ■

II.4.19 Example By Example II.4.17, the theorem is obviously not true in the case $n = 0$, r arbitrary, if we drop the quasistrongly graded hypothesis. In view of some applications in Chapter V, it is natural to ask whether it holds for $r = 0$. We give a counterexample.

Let R be the ring of the preceding example, gr-semilocalized at p and q, and let $S = R_p^g \times R_q^g$. As we have seen earlier, $R_p^g \otimes R_q^g$ is just the gr-localization of R at the graded prime ideal which is maximal among those contained in $p \cap q$, so $R_p^g \otimes R_q^g = K$, the graded field of fractions of R. As $\text{char}(K) = 0$, we have that $L = K[X]/(X^p - T)$ is a gr-étale covering of K. Clearly $S \otimes S = R_p^g \otimes R_p^g \times R_p^g \otimes R_q^g \times R_q^g \otimes R_p^g \times R_q^g \otimes R_q^g = R_p^g \times K \times K \times R_q^g$ and $T = R_p^g \times L \times L \times R_q^g$ is a gr-étale covering of $S \times S$. Now we claim that there is no gr-étale covering $S \to R'$ such that $S^{(2)} \to S'^{(2)}$ factorizes through T.

for this r we get $a_{er} = d(er, i(r))(d(-er, i(r))a_{er}) \in D_{er}A_0 \subset P_+$. We thus have established that for any $n(0) \in \mathbf{N}$, we may find $r \in \mathbf{N}$ such that $r \geq n(0)$ and $d(er, i(r)) \notin P_+$ for some $d(er, i(r))$. The element thus selected will be written as d_{er}. Define P_{-k} to consist of all $x \in A_{-k}$ such that $xy_{k+n} \in P_n$ for all $n \geq 0$ and all $y_{k+n} \in A_{k+n}$. It is clear that P_{-k} is an A_0-module, so we may put $P_- = \oplus_{k>0} P_{-k}$ and $P = P_- \oplus P_+$ is a two-sided A_0-module. Let us check that P is a left ideal of A, by showing that for any $z_h \in A_h$ and $x_k \in P_k$ $(h, k \in \mathbf{Z})$ we have that $z_h x_k \in P_{h+k}$. We have to distinguish several cases.

(4.3.1) $h \geq 0$ *and* $k \geq 0$: Then copy, since P_+ is an ideal of A_0.

(4.3.2) $h < 0$, $k \geq 0$, *and* $h + k \geq 0$: Choose r large enough so that $d_{er} \notin P_+$ and $er + h \geq 0$; then $(d_{er} z_h) x_k \in P_+$, but as $z_h x_k \in A_{>0}$ and $d_{er} \notin P_+$ is central, it follows that $z_h x_k \in P_{h+k}$.

(4.3.3) $h < 0$, $k \geq 0$ *and* $h + k < 0$: To show that for all $n \geq 0$ and all $y_{n-h-k} \in A_{n-h-k}$ we have $z_h x_k y_{n-h-k} \in P_n$, note that $n - h - k > 0$, $x_k y_{n-h-k} \in P_{n-h}$, so $z_h(x_k y_{n-h-k}) \in P_n$, by (4.3.2).

(4.3.4) $h \geq 0$, $k < 0$ *and* $h + k \geq 0$: Choose r large enough so that $d_{er} \notin P_+$ and $er + k \geq 0$; then $z_h(x_k d_{er})$ has the property that $x_k d_{er} \in P_+$, but then $z_h x_k d_{er} \in P_+$, by (4.3.1), hence $z_h x_k \in P_+$.

(4.3.5) $h \geq 0$, $k < 0$, *and* $h + k < 0$: Consider $n \geq 0$ such that $n - h - k \geq 0$ and $z_h x_k y_{n-h-k} \in P_n$ for any $y_{n-h-k} \in A_{n-h-k}$; then the definition of P_k yields $x_k y_{n-h-k} \in P_{n-h}$ and $z_h x_k y_{n-h-k} \in P_n$ by (4.3.1).

(4.3.6) $h < 0$ *and* $k < 0$: Look at $y_{n-h-k} \in A_{n-h-k}$ for $n \geq 0$; then $z_h(x_k y_{n-h-k}) \in P_n$, because $x_k y_{n-h-k} \in P_{n-h}$ by definition. It follows that $AP_+ \subset P_- \oplus P_+ = P$. On the other hand, if $x_k \in P_k$ for some $k < 0$, then choosing r large enough such that $re + k \geq 0$ and $d_{er} \notin P_+$, yields $x_k = x_k \cdot 1 = x_k \sum_r d(re, i(r)) d(-re, i(r)) \in AP_+$ and similarly $x_k \in P_+ A$; hence $AP_+ = P$ and $P_+ A \subset P$, hence $PA = AP_+ A \subset AP = A$, so clearly $PA = P$ and $AP_+ A = AP_+$.

Now, suppose that $x_k A y_h \in P$ for some $x_k \in A_k$ and $y_h \in A_h$. Choose r such that $d_{er} \notin P_+$ and $er + k \geq 0$, $er + h \geq 0$; then $d_{er} x_k A d_{er} y_h \in P$ yields that $(d_{er} x_k) A_{\geq 0} (d_{er} y_h) \in P_+$, hence $d_{er} x_k \in P_+$, say $d_{er} x_k \in P_+$. If $k \geq 0$, it follows that $x_k \in P_+$. If $k < 0$, then for any $y - k + n$ in A_{n-k}, with $n \geq 0$, we have $d_{er} x_k y_{n-k} \in P_+$ and since $d_{er} \notin P_+$, it follows that $x_k y_{n-k} \in P_+$. By definition $x_k \in P_{-k}$. Hence P is a prime ideal of A such that $P \cap A_0 = P_+$. If P and Q are prime ideals of A such that $P_+ = Q_+$ and $P_{-k} \not\subset Q_{-k}$ for some $k > 0$, then from $P_{-k} A_{k+n} \subset P_n = Q_n$, it follows that $P_{-k} A_{k+n} A \subset Q$ for $n \geq 0$. But $A A_{k+n} = A$ holds for any n, such that $k + n = er$, for example. Therefore $P_{-k} \subset Q$ contradicts $P_{-k} \not\subset Q_{-k}$, that is, it follows that $P = Q$ if $P_+ = Q_+$. This proves the result. ∎

III.4.4 Remarks (4.4.1) Both $\mathrm{Spec}^g(A)$ and $\mathrm{Proj}(A_{\geq 0})$ are endowed with the Zariski topologies. One easily checks that the above bijection actually yields a homeomorphism.

(4.4.2) If $P \in \mathrm{Spec}^g(A)$, then $A/P \cong_g M_n(K[X, X^{-1}, \varphi])(\underline{d})$ for some skewfield K, and some variable X of degree s/t such that $X\lambda = \lambda^{\varphi} X$ for all $\lambda \in K$, where φ is an automorphism of K such that $\varphi^{t/s}$ is an inner automorphism of K and where $\underline{d} \in \mathbf{Z}^n$ describes the gradation on A/P. On the other hand, $A_{\geq 0}/P_+ = (A/P)_{\geq 0} \cong_g M_n(K[X, \varphi])(\underline{d})$. In particular, the *PI* degree of P and of P_+ are equal. It also follows that $A_{\geq 0}$ need not be an Azumaya algebra—one should call it a *projective Azumaya algebra*.

(4.4.3) It is easy to check that $Z(A_{\geq 0}) = Z(A)_{\geq 0}$. Indeed, if z centralizes $A_{\geq 0}$ and $y \in A_{-k}$ for some $k > 0$, then $y = \sum_{i(r)} (y d(re, i(r))) d(-re, i(r))$ for some $r \in \mathbf{N}$ such that $re - k \geq 0$; hence z commutes with y as well.

III.4.5 We now come to the definition of the graded Brauer group. Let $\underline{\underline{Az}}_g(R)$ be the category of graded R-Azumaya algebras, with graded R-algebra homomorphisms. The tensor product over R is a product for this category. The functor $\mathrm{END}_R : \underline{\underline{FP}}_g(R) \to \underline{\underline{Az}}_g(R)$ induces a map $K_0 \underline{\underline{FP}}_g(R) \to K_0 \underline{\underline{Az}}_g(R)$. We define the *graded Brauer group* of R, denoted by $\mathrm{Br}_g(R)$, to be the cokernel of $K_0 \mathrm{END}_R$:

$$\mathrm{Br}_g(R) = \mathrm{coker}(K_0 \mathrm{END}_R : K_0 \underline{\underline{FP}}_g(R) \longrightarrow K_0 \underline{\underline{Az}}_g(R)).$$

Of course, $\mathrm{Br}_g(R)$ may also be described as follows. Call two graded R-Azumaya algebras A and B *gr-equivalent*, if there exists a couple of graded faithfully projective R-modules P, Q such that the graded R-algebras $A \otimes_R \mathrm{END}_R(P)$ and $B \otimes_R \mathrm{END}_R(Q)$ are gr-isomorphic. The set of gr-equivalence classes $[A]$ of graded R-Azumaya algebras is endowed with a group structure by defining $[A] \cdot [B] = [A \otimes_R B]$. Moreover, since $\mathrm{END}_R(A) \cong_g A \otimes_R A^{\mathrm{opp}}$, clearly $[A]^{-1} = [A^{\mathrm{opp}}]$.

III.4.6 Proposition For any graded Azumaya algebra A, we have $[A] = [R]$ if and only if $A \cong_g \mathrm{END}_R(E)$ for some graded faithfully projective R-module E.

Proof: By definition, we have $[A] = [R]$ if and only if we may find graded faithfully projective R-modules P and Q such that $A \otimes_R \mathrm{END}_R(P) \cong_g \mathrm{END}_R(Q)$. Put $B = \mathrm{END}_R(P)$; then A and B may be identified with subalgebras of $\mathrm{END}_R(Q)$ and since the product map $A \otimes_R B \to \mathrm{END}_R(Q)$ is an isomorphism, it follows that $\mathrm{END}_R(Q)^A = \mathrm{End}_R(Q)^A = B$ and $\mathrm{END}_R(Q)^B = \mathrm{End}_R(Q)^B = A$, by the double commutator theorem (cf. [37], for example). Now, Q may be viewed as a left B-module in the obvious way and it is clear that the R-module $E = P^* \otimes_B Q$ corresponds to Q under the category equivalence between R-gr and B-gr by the Morita theorems. Hence $\mathrm{END}_R(M) \cong_g \mathrm{END}_B(Q) = \mathrm{END}_R(Q)^B = A$ and it suffices to check that E is a faithfully projective graded R-module to finish the proof. Again, as in the nongraded case, this follows easily from the Morita theorems applied to the category equivalence R-gr $\to B$-gr induced by P. ∎

III.4.7 Corollary If A and B are graded Azumaya algebras, then $[A] = [B]$ in $\mathrm{Br}_g(R)$ if and only if $A \otimes_R B^{\mathrm{opp}} \cong_g \mathrm{END}_R(P)$ for some $P \in \underline{\underline{FP}}_g(R)$. ∎

III.4.8 It is clear that Br_g is actually a covariant functor from the category of commutative graded rings (with graded ring homomorphisms) to the category of abelian groups. If $f : R \to S$ is a morphism of graded rings, then the induced morphism on graded Brauer groups is the morphism $\mathrm{Br}_g(f) : \mathrm{Br}_g(R) \to \mathrm{Br}_g(S)$ which sends $[A]$ to $[S \otimes_R A]$. One easily checks this to be well defined—note that one has to use the fact that $S \otimes_R P \in \underline{\underline{FP}}_g(S)$ for every $P \in \underline{\underline{FP}}_g(R)$, and that in this case $S \otimes_R \mathrm{END}_R(P) \cong_g \mathrm{END}_S(S \otimes_R P)$.

We say that S (or f) *splits* a graded R-Azumaya algebra A if $[A] \in \mathrm{Br}_g(R)$ lies in the kernel of $\mathrm{Br}_g(f)$, that is, if $S \otimes_R A \cong_g \mathrm{END}_S(Q)$ for some graded S-progenerator Q. We will denote $\mathrm{Ker}(\mathrm{Br}_g(f))$ by $\mathrm{Br}_g(S/R)$.

III.4.9 Proposition If R is strongly graded, then $\mathrm{Br}_g(R) \cong \mathrm{Br}(R_0)$.

Proof: Since R is strongly graded, every graded Azumaya algebra A over R is a strongly graded ring, and so will be A^e. By the equivalence of the categories $(A^e)_0$-mod and A^e-gr, A will be a projective A^e-module if and only if A_0 is a projective $(A^e)_0$-module. For every $n \in \mathbf{Z}$, $R_{-n} \otimes_{R_0} R_n = R_{-n}R_n = R_0$, so it follows that $A_n \otimes_{R_0} A_{-n}^{\mathrm{opp}} = A_nR_{-n} \otimes_{R_0} R_nA_{-n}^{\mathrm{opp}}$. Hence $(A^e)_0 = \sum_n A_n \otimes_{R_0} A_{-n} = A_0 \otimes_{R_0} A_0^{\mathrm{opp}} = (A_0)^e$. The maps $\mathrm{Br}_g(R) \to \mathrm{Br}(R_0) : [A] \to [A_0]$ and $\mathrm{Br}(R_0) \to \mathrm{Br}_g(R) : [A] \to [A \otimes_{R_0} R]$ are inverse to each other; hence the statement follows.

Note that taking parts of degree zero (or conversely, $\otimes_{R_0} R$) respects the equivalence relations defining the Brauer groups involved! ■

Our next aim is to deduce graded versions of Bass' exact sequence [7, III.6].

III.4.10 Theorem (First graded version of Bass' exact sequence) Let R be a commutative graded ring for which $\delta(R) < +\infty$ and consider the sequence of functors

$$\underline{\underline{\mathrm{Pic}}}_g(R) \xrightarrow{I} \underline{\underline{FP}}_g(R) \xrightarrow{\mathrm{END}} \underline{\underline{Az}}_g(R)$$

of categories with product; they induce the following exact sequence:

$$U_0(R) = K_1 \underline{\underline{\mathrm{Pic}}}_g(R) \xrightarrow{K_1 I} K_1 \underline{\underline{FP}}_g(R) \xrightarrow{K_1 \mathrm{END}} K_1 \underline{\underline{Az}}_g(R)$$
$$\longrightarrow \mathrm{Pic}_g(R) = K_0 \underline{\underline{\mathrm{Pic}}}_g(R) \xrightarrow{K_0 I} K_0 \underline{\underline{FP}}_g(R)$$
$$\xrightarrow{K_0 \mathrm{END}} K_0 \underline{\underline{Az}}_g(R) \longrightarrow \mathrm{Br}_g(R) \longrightarrow 1$$

Furthermore, $\mathrm{Ker}(K_1 I) = U_0(R)_{\mathrm{tors}}$ and $\mathrm{Ker}(K_0 I) = \mathrm{Pic}_g(R)_{\mathrm{tors}}$.

Proof: By definition, $\mathrm{Br}_g(R) = \mathrm{Coker}(K_0 \mathrm{END})$. As END is a cofinal preserving functor, we have an exact sequence (cf. [6, VII.2])

$$K_1 \underline{\underline{FP}}_g(R) \longrightarrow K_1 \underline{\underline{Az}}_g(R) \longrightarrow K_1 \Phi\, \mathrm{END}$$
$$\longrightarrow K_0 \underline{\underline{FP}}_g(R) \longrightarrow K_0 \underline{\underline{Az}}_g(R)$$

Let us prove the following lemma:

III.4.11 Lemma $K_1 \Phi\, \mathrm{END} \cong \mathrm{Pic}_g(R)$.

Proof: Recall that the elements of $K_1 \Phi\, \mathrm{END}$ are generated by objects of the form (P, α, Q), where $P, Q \in \underline{\underline{FP}}_g(R)$, and α is a graded isomorphism $\mathrm{END}_R(P) \to \mathrm{END}_R(Q)$. Now take $[L] \in \underline{\underline{\mathrm{Pic}}}_g(R)$, and set $T([L]) = [(L, \alpha_L, R)]$, where α_L is the unique graded R-algebra isomorphism $\mathrm{END}_R(L) \cong_g R \to \mathrm{END}_R(R) \cong_g R$.

Conversely, let (P,α,Q) represent an element of $K_1\Phi\,\mathrm{END}$. Then $\alpha : A = \mathrm{END}_R(P) \to B = \mathrm{END}_R(Q)$ allows us to view graded left B-modules as graded left A-modules. By the graded version of the Morita theorems, $P\otimes_R - : R\text{-gr} \to A\text{-gr}$ is a category equivalence, the inverse being $\mathrm{HOM}_A(P,\_)$. Apply this to the graded B-module (and hence A-module) Q to obtain a graded R-module $L = \mathrm{HOM}_R(P,Q)$. As $Q \cong_g P \otimes_R L$, it follows that $L \in \underline{\underline{\mathrm{Pic}}}_g(R)$. We define $S([(P,\alpha,Q)]) = [L]$. We leave it to the reader to verify that S and T are well defined group homomorphisms, and that $T\circ S = 1$, $S \circ T = 1$. The details are just modifications of [7, III.6.5]. ■

III.4.12 Lemma $\mathrm{Ker}(K_0 I) = \mathrm{Pic}_g(R)_{\mathrm{tors}}$.

Proof: Suppose that $[J] \in \mathrm{Pic}_g(R)_{\mathrm{tors}}$, so $J^{(n)} \cong_g R$, for some integer n. Set $E = R \oplus J \oplus J^{(2)} \oplus \cdots \oplus J^{(n-1)} \in \underline{\underline{FP}}_g(R)$; then $E \otimes_R J \cong_g E$. Also, by III.3.18, there exists $F \in \underline{\underline{FP}}_g(R)$ such that $E \otimes_R F \cong_g R^m(\underline{d})$. Therefore we have that $J^m(\underline{d}) \cong_g J \otimes_R R^m(\underline{d}) \cong_g J \otimes_R E \otimes_R F \cong_g E \otimes_R F \cong_g R^m(\underline{d})$. Hence $[J] = [R] = 1$ in $K_0\,\underline{\underline{FP}}_g(R)$. The converse implication, that is,if $[J] \in \mathrm{Pic}_g(R)$ vanishes in $K_0\,\underline{\underline{FP}}_g(R)$, then $[J] \in \mathrm{Pic}_g(R)_{\mathrm{tors}}$, may be deduced in exactly the same way. ■

It is clear that $K_1\mathrm{END} \circ K_1 I = 1$. So we still have to show that $\mathrm{Ker}(K_1\mathrm{END}) \subset \mathrm{Im}(K_1 I)$. First, we need some preliminary results. We introduce the index set

$$\Gamma = \{(n,\underline{d}) = \underline{d} : n \in \mathbf{N}_0, \underline{d} \in \mathbf{Z}^n\},$$

on which a multiplication rule is defined by

$$(n,\underline{d}) \cdot (m,\underline{e}) = (nm,\underline{f}),$$

where $f_{\alpha+\beta n} = d_\alpha + e_\beta$, for $1 \le \alpha \le n$, $0 \le \beta \le m-1$. Since the full subcategory of $\underline{\underline{FP}}_g(R)$ consisting of all graded free R-

modules $R^n(\underline{d})$, and the full subcategory of $\underline{\underline{Az}}_g(R)$ consisting of the matrix rings $M_n(R)(\underline{d})$ are cofinal, we have by [6, VII.2.3] that

$$K_1 \underline{\underline{FP}}_g(R) = \varinjlim W_n^0(\underline{d})$$
$$K_1 \underline{\underline{Az}}_g(R) = \varinjlim V_n^0(\underline{d})$$

where

$$W_n^\circ(\underline{d}) = Gl^0(\underline{d})/[Gl^0(\underline{d}), Gl^0(\underline{d})]$$
$$V_n^0(\underline{d}) = PGl^0(\underline{d})/[PGl^0(\underline{d}), PGl^0(\underline{d})]$$

and

$$Gl^0(\underline{d}) = Gl_n^0(R)(\underline{d}) = \mathrm{Aut}_0(R^n(\underline{d}))$$
$$PGl^0(\underline{d}) = PGl_n^0(R)(\underline{d}) = \mathrm{Aut}_0(M_n(R)(\underline{d})).$$

The inductive limits above are taken over Γ, ordered by divisibility. By $E_n^0(R)(\underline{d})$, we mean the subgroup of $Gl_n^0(R)(\underline{d})$ generated by all elementary matrices of degree zero, that is, by the matrices of the form $I + ae_{ij}$, where $i \neq j$, and where $\deg a = d_i - d_j$.

III.4.13 Lemma Suppose that every index occurring in the multiindex $\underline{d}$ occurs at least three times in it. Then $E_n^0(R)(\underline{d}) \subset [E_n^0(R)(\underline{d}), E_n^0(R)(\underline{d})]$.

Proof: Take an elementary matrix $I + ae_{ik}$, and let j be an index different from i and k such that $d_i = d_j$. Then it is well known that $[I + e_{ij}, I + ae_{jk}] = I + ae_{ik}$ (cf. [6, V.1.2.b]). Also $\deg(I + e_{ij}) = \deg(I + ae_{jk}) = 0$. ■

III.4.14 Lemma Let $M \in M_n(R)(\underline{d})$ be a matrix of degree zero of which all entries are integers, and suppose that $\det M = 1$. Then $M \in E_n^0(R)(\underline{d})$.

Proof: Furnish $\mathbf{Z}$ with the trivial gradation, and take $\alpha = (a_1, \ldots, a_n) \in h(\mathbf{Z}^n(\underline{d}))$. By the proof of [6, IV.5.9], there exists $\varepsilon \in E_n(R)$ such that $\varepsilon\alpha = (a, 0, \ldots, 0)$. Now ε may be chosen to be homogeneous of degree zero. Indeed, suppose $\deg \alpha = d$. Then all a_i are zero, except for those i satisfying $d_i = d$. So we may suppose that $\underline{d}$ is constant. But in this case, $M^n(\mathbf{Z})(\underline{d})_0 = M_n(\mathbf{Z})$. The result follows easily now, as in [6]. ■

III.4.15 Lemma $\mathrm{Ker}(K_1\mathrm{END}) \subset \mathrm{Im}(K_1 I)$.

Proof: For $\underline{\underline{d}} \in \Gamma$, consider $f(\underline{\underline{d}}) : Gl_n^0(R)(\underline{\underline{d}}) \to PGl_n^0(R)(\underline{\underline{d}})$, taking inner automorphisms. Then $\mathrm{Ker}(f(\underline{\underline{d}})) = U_0(R)$, and $\mathrm{Im}(f(\underline{\underline{d}})) = \mathrm{Inn}\, PGl_n^0(R)(\underline{\underline{d}}) = \mathrm{Inn}(\underline{\underline{d}})$, say. So, $\mathrm{Coker}(K_1 I) = \mathrm{Inn}(\underline{\underline{d}})/[\mathrm{Inn}(\underline{\underline{d}}), \mathrm{Inn}(\underline{\underline{d}})]$, and we are done if we can show that the inclusion $\mathrm{Inn}(\underline{\underline{d}}) \to PGl(\underline{\underline{d}})$ induces a monomorphism $\mathrm{Inn}(\underline{\underline{d}})/[\mathrm{Inn}(\underline{\underline{d}}), \mathrm{Inn}(\underline{\underline{d}})] \to PGl(\underline{\underline{d}})/[PGl(\underline{\underline{d}}), PGl(\underline{\underline{d}})]$, or, equivalently, if $[PGl(\underline{\underline{d}}), PGl(\underline{\underline{d}})] \subset [\mathrm{Inn}(\underline{\underline{d}}), \mathrm{Inn}(\underline{\underline{d}})]$. Take $\alpha, \beta \in PGl_n^0(R)(\underline{\underline{d}}) = PGl(\underline{\underline{d}})$, and let $\tau : R^n(\underline{\underline{d}}) \otimes R^n(\underline{\underline{d}}) \to R^n(\underline{\underline{d}}) \otimes R^n(\underline{\underline{d}})$ be the switch map. Write $E(\tau)$ for the induced inner automorphism of $M_n(R)(\underline{\underline{d}} \cdot \underline{\underline{d}})$. Then $E(\tau)(\alpha \otimes 1)E(\tau)^{-1} = 1 \otimes \alpha$ commutes with $\beta \otimes 1$. Now the entries of the matrix of τ are integers, and $\deg \tau = 0$. Therefore, if $n(n-1)$ is divisible by 4, for example, if n is divisible by 4, then $\tau \in E_n^0(R)(\underline{\underline{d}}) \subset [E_n^0(R)(\underline{\underline{d}}), E_n^0(R)(\underline{\underline{d}})] \subset [Gl(\underline{\underline{d}}), Gl(\underline{\underline{d}})]$, if we suppose furthermore that $\underline{d}$ satisfies the condition of III.4.13. It follows that the image of $PGl(\underline{\underline{d}})$ in $PGl(\underline{\underline{d}} \cdot \underline{\underline{d}})/[\mathrm{Inn}(\underline{\underline{d}} \cdot \underline{\underline{d}}), \mathrm{Inn}(\underline{\underline{d}}, \underline{\underline{d}})]$ is abelian. Observe that this factor group is well defined, because $\mathrm{Inn}(\underline{\underline{d}} \cdot \underline{\underline{d}})$ and its commutator are normal in $PGl(\underline{\underline{d}} \cdot \underline{\underline{d}})$. Using the fact that $\{\underline{\underline{d}} = (n, \underline{d}) : \underline{d}$ satisfies the condition of III.4.13 and n is divisible by 4$\}$ is a cofinal subset of Γ, we obtain the result after passing to the limits. ■

III.4.16 Lemma $\mathrm{Ker}(K_1 I) = U_0(R)_{\mathrm{tors}}$.

Proof: Take $a \in U_0(R)$ such that $a^n = 1$. Represent $K_1 I(a)$ by the diagonal matrix $aI \in Gl_n(R)(\underline{0})$. Then, by Whitehead's lemma ([6, V.1.7]), we have

$$\begin{bmatrix} a & 0 & \dots & 0 \\ 0 & a & \dots & 0 \\ \dots & & & \\ 0 & 0 & \dots & a \end{bmatrix} \equiv \begin{bmatrix} a^n & 0 & \dots & 0 \\ 0 & 1 & \dots & 0 \\ \dots & & & \\ 0 & 0 & \dots & 1 \end{bmatrix} \equiv I \operatorname{mod} E_n^0(R)(\underline{0})$$

If we take $n \geq 3$, then we may conclude that $K_1 I(a) = 1$, using III.4.13. Conversely, suppose that for some $\underline{d} \in \mathbf{Z}^n$, $aI \in [Gl_n^0(R)(\underline{d}), Gl_n^0(R)(\underline{d})]$. Then $aI \in [Gl_n(R), Gl_n(R)]$, so the image of $a \in U_0(R) \subset U(R)$ in $K_1 \underline{\underline{FP}}(R)$ is trivial. By Bass' result [7, III.6.8], it follows that $a^n = 1$. This finishes the proof of Lemma III.4.16 and Theorem III.4.10. ■ ■

III.4.17 Caution The graded Brauer group was originally introduced by the second author in [61]. The notation $\mathrm{Br}^g(R)$ was used there. Of course, this notation is not consistent with our conventions about upper and lower index g. $\mathrm{Br}^g(R)$ should denote $\mathrm{Coker}(K_0 FP^g(R) \to K_0 Az^g(R))$.

III.4.18 Theorem (Second graded version of Bass' exact sequence) Let R be a commutative graded ring, and consider the following commutative diagram of functors of categories with product:

$$\begin{array}{ccccc} \underline{\underline{\mathrm{Pic}}}^g(R) & \xrightarrow{I^g} & \underline{\underline{FP}}^g(R) & \xrightarrow{\mathrm{END}^g} & \underline{\underline{Az}}^g(R) \\ \downarrow U_0 & & \downarrow U_1 & & \downarrow U_2 \\ \underline{\underline{\mathrm{Pic}}}(R) & \xrightarrow{I} & \underline{\underline{FP}}(R) & \xrightarrow{\mathrm{END}} & \underline{\underline{Az}}(R) \end{array}$$

It induces the following commutative diagram of exact sequences:

$$U(R) = K_1 \underline{\underline{\mathrm{Pic}}}^g(R) = K_1 \underline{\underline{\mathrm{Pic}}}(R) \longrightarrow K_1 \underline{\underline{FP}}^g(R)$$
$$= K_1 \underline{\underline{FP}}(R) \longrightarrow K_1 \underline{\underline{Az}}^g(R) = K^1 \underline{\underline{Az}}(R)$$

$$\begin{array}{l} \nearrow \mathrm{Pic}^g(R) \xrightarrow{K_0 I^g} K_0 \underline{\underline{FP}}^g(R) \xrightarrow{K_0\,\mathrm{END}^g} K_0 \underline{\underline{Az}}^g(R) \longrightarrow \mathrm{Br}^g(R) \longrightarrow 1 \\ \searrow \mathrm{Pic}(R) \xrightarrow{K_0 I} K_0 \underline{\underline{FP}}(R) \xrightarrow{K_0\,\mathrm{END}} K_0 \underline{\underline{Az}}(R) \longrightarrow \mathrm{Br}(R) \longrightarrow 1 \end{array}$$

By definition, $\mathrm{Br}^g(R) = K_0\mathrm{END}$. Also, $\mathrm{Ker}(K_1 I) = U(R)_{\mathrm{tors}}$, $\mathrm{Ker}(K_0 I^g) = \mathrm{Pic}^g(R)_{\mathrm{tors}}$, and $\mathrm{Ker}(K_0 I) = \mathrm{Pic}(R)_{\mathrm{tors}}$.

Proof: The bottom row is just Bass' exact sequence. The isomorphisms of the K_1-groups are also clear, since the functors U_i are product preserving and cofinal. As in III.4.10, there is no problem in proving that $K_1 \Phi\,\mathrm{END}^g = \mathrm{Pic}^g(R)$. It has been proved in an earlier section that $K_0 \underline{\underline{FP}}^g(R) \to K_0 \underline{\underline{FP}}(R)$ is a monomorphism, and this proof goes through for the homomorphism $K_0 \underline{\underline{Az}}^g(R) \to K_0 \underline{\underline{Az}}(R)$. Two of the three last equalities are results of Bass; the third follows by restriction. ■

III.4.19 We now present an analog of the so-called Brauer class group. Consider the category with objects (E, α, A), where $E \in \underline{\underline{FP}}(R)$, $A \in \underline{\underline{Az}}_g(R)$, and α an automorphism $\mathrm{End}_R(E) \to \underline{A}$ in $\underline{\underline{Az}}(R)$. A homomorphism $(E, \alpha, A) \to (E', \alpha', A')$ will be a pair (f, φ), where $f : E \to E'$ is a morphism in $\underline{\underline{FP}}(R)$, and $\varphi : A \to A'$ in $\underline{\underline{Az}}_g(R)$ such that $\underline{\varphi} \circ \alpha = \alpha = \alpha' \circ \mathrm{End} f$. Now take the set of equivalence classes Ω on this category, and put the following equivalence relation on it:

$(E, \alpha, A) \sim (E', \alpha', A')$ if and only if there exist $P, Q \in \underline{\underline{FP}}_g(R)$ such that $(E \otimes P, \alpha \otimes Id, A \otimes \mathrm{End}_R(P)) \cong (E' \otimes Q, \alpha' \otimes Id, A' \otimes \mathrm{End}_R(Q))$.

We define $GR(R) = \Omega/\sim$.

III.4.20 Proposition We have an exact sequence

$$\mathrm{Pic}_g(R) \xrightarrow{f_0} \mathrm{Pic}(R) \xrightarrow{f_1} GR(R) \xrightarrow{f_2} \mathrm{Br}_g(R) \xrightarrow{f_3} \mathrm{Br}(R)$$

Proof: For $I \in \underline{\underline{\mathrm{Pic}}}(R)$, let α be an isomorphism $\mathrm{End}_R(I) \to R$. We define $f_1([I]) = [(I, \alpha, R)]$. $f_1([I])$ is independent of the choice of α, because $[(I,\alpha,R)][(I,\beta,R)]^{-1} = [(I \otimes I^*, \alpha \otimes \beta^*, R)] = [(R,\alpha \otimes \beta^*, R)] = 1$ in $GR(R)$. Now define $f_2([E,\alpha,A)]) = [A]$. It is easy to verify that f_1 and f_2 are well defined homomorphisms. Clearly $f_1 \circ f_0 = 1$, $f_2 \circ f_1 = 1$, $f_3 \circ f_2 = 1$. Suppose $f_1([I]) = 1$. Then $(I,\alpha,R) \cong (P,Id,\mathrm{End}_R(P))$, so $[I] = [P] \in \mathrm{Im}(f_0)$. If $f_2[(E,\alpha,A)] = 1$, then $A = \mathrm{END}_R(P)$ for some $P \in \underline{\underline{FP}}_g(R)$. E is an A-module by the action $a \cdot x = \alpha^{-1}(a)(x)$. Furthermore $\mathrm{Hom}_A(\underline{P}, E) \otimes \underline{P} \cong (\underline{P}^* \otimes_A E) \otimes \underline{P} \cong (E \otimes_{A^{\mathrm{opp}}} \underline{P}^*) \otimes \underline{P} \cong E \otimes_{A^{\mathrm{opp}}} \mathrm{End}_R(\underline{P}) \cong A \otimes_A E \cong E$. Let $F = \mathrm{Hom}_A(\underline{P}, E)$; then $[(E,\alpha,A)] = [(F \otimes P, 1 \otimes \alpha, R \otimes A)] = [(F,1,R)] = f_1([F])$. Clearly $F \in \underline{\underline{\mathrm{Pic}}}(R)$. Finally, suppose $f_3([A]) = 1$; then $A \cong \mathrm{End}_R(E)$. Then $[A] = f_2([(E,\alpha,A)])$. ∎

III.5 GRADED COHOMOLOGY GROUPS AND THE CROSSED PRODUCT THEOREMS

III.5.1 Let us start from a pair of complexes of abelian groups (A_1,d_1) and (A_2,d_2) and let $f : (A_1,d_1) \to (A_2,d_2)$ be a morphism of complexes of degree zero, that is, we have a commutative diagram

$$\begin{array}{ccccccccc}
\cdots & \longrightarrow & A_1^{n-1} & \xrightarrow{d_1^{n-1}} & A_1^n & \xrightarrow{d_1^n} & A_1^{n+1} & \longrightarrow & \cdots \\
 & & \downarrow f^{n-1} & & \downarrow f^n & & \downarrow f^{n+1} & & \\
\cdots & \longrightarrow & A_2^{n-1} & \xrightarrow{d_2^{n-1}} & A_2^n & \xrightarrow{d_2^n} & A_2^{n+1} & \longrightarrow & \cdots
\end{array}$$

We define a new complex (A,d) by $A^n = A_1^n \times A_2^{n-1}$ and $d^n : A^n \to A^{n+1}$ given by $d^n(a_1,a_2) = (d_1^n(a_1), d_2^{n-1}(a_2) - f_n(a_1))$ for any $a_1 \in A_1^n$, $a_2 \in A_2^{n-1}$. One easily checks that (A,d) is a complex. The corresponding cohomology groups are denoted by $H^n_{A_2}(A_1) = H^n(A) = \mathrm{Ker}(d^n)/\mathrm{Im}(d^{n-1})$.

III.5.2 Proposition There exists a long exact sequence of abelian groups:

$$\cdots \longrightarrow H^{n-1}(A_2) \longrightarrow H^n_{A_2}(A_1)$$
$$\longrightarrow H^n(A_1) \longrightarrow H^n(A_2) \longrightarrow \cdots$$

Proof: The exact sequences $0 \to A_2^{n-1} \to A^n \to A_1^n \to 0$ induce an exact sequence of complexes $0 \to (A_2(-1), d_2(-1)) \to (A, d) \to (A_1, d_1) \to 0$, where $(A_2(-1), d_2(-1))$ is the complex (A_2, d_2) shifted by -1, and then the above exact sequence is just the long cohomology sequence, taking into account the fact that $H^{n-1}(A_2) = H^n(A_2(-1))$.

III.5.3 We shall use this tool in the following cases.
(5.3.1) Let S be a commutative graded R-algebra and let A_1 and A_2 be the Amitsur complexes associated to the functors U and gr, respectively, that is, $A_1 = \mathcal{C}(S/R, U)$ [resp., $A_2 = \mathcal{C}(S/R, \mathrm{gr})$] and $f^{n-1} : U(S^{(n-1)}) \to \mathrm{gr}(S^{(n-1)})$ is defined as in the preceding section. We then obtain the gr-cohomology groups $H^n_{\mathbf{gr}}(S/R, U) = H^n_{A_2}(A_1)$ of U.
(5.3.2) If S is a graded Galois extension of R with Galois group $G = \mathrm{Gal}(S/R)$ and if $K(G, U(S))$ [resp., $K(G, \mathrm{gr}(S))$] are the complexes associated to the G-$\mathbf{Z}$-modules $U(S)$ and $\mathrm{gr}(S)$, then we are led to the definition of $H^n_{\mathbf{gr}}(G, U(S))$. Using previous results, together with the techniques developed (in the ungraded case) in [18, 37], one easily verifies that there is an isomorphism $H^n_{\mathbf{gr}}(G, U(S)) \cong H^n_{\mathbf{gr}}(S/R, U)$.

III.5.4 Let us assume throughout that S is a commutative graded R-algebra. The Amitsur complexes of U and gr are denoted by

$$\begin{array}{ccccccccc}
1 & \longrightarrow & U(S) & \xrightarrow{\Delta_0} & U(S^{(2)}) & \xrightarrow{\Delta_1} & U(S^{(3)}) & \xrightarrow{\Delta_2} & \cdots \\
 & & \downarrow{\scriptstyle d_0} & & \downarrow{\scriptstyle d_1} & & \downarrow{\scriptstyle d_2} & & \\
1 & \longrightarrow & \mathrm{gr}(S) & \xrightarrow{D_0} & \mathrm{gr}(S^{(2)}) & \xrightarrow{D_1} & \mathrm{gr}(S^{(3)}) & \xrightarrow{D_2} & \cdots
\end{array}$$

III.5.5 Theorem (First Graded Version of the Chase–Rosenberg Exact Sequence) Let R be a commutative graded ring, and S a commutative graded R-algebra which is faithfully projective as an R-module. Then we have an exact sequence

$$\begin{aligned}
1 &\longrightarrow H^0(S/R,U_0) \xrightarrow{\alpha_0} U_0(R) \longrightarrow 1 \\
&\longrightarrow H^1(S/R,U_0) \xrightarrow{\alpha_1} \mathrm{Pic}_g(R) \xrightarrow{\beta_1} H^0(S/R,\mathrm{Pic}_g) \\
&\xrightarrow{\gamma_1} H^2(S/R,U_0) \xrightarrow{\alpha_2} \mathrm{Br}_g(S/R) \xrightarrow{\beta_2} H^1(S/R,\mathrm{Pic}_g) \\
&\xrightarrow{\gamma_2} H^3(S/R,U_0).
\end{aligned}$$

Proof: The most direct way to prove this theorem is to adapt the proof of the exactness of the classical Chase–Rosenberg sequence given by Knus in [36]. As there is no problem in doing this, we content ourselves by giving a description of the connecting maps.

(5.5.1) *Definition of* α_0: We use the theorem of faithfully flat descent of elements: If $u \in U_0(S)$ represents an element of $H^0(S/R,U)$, then $u \otimes 1 = 1 \otimes u$, so $u \in U_0(R)$.

(5.5.2) *Definition of* α_1: If $t \in U_0(S^{(2)})$ is a cocycle, then multiplication by t is a graded descent datum, defining an invertible graded R-module I. We define $\alpha_1([T]) = [I]$.

(5.5.3) *Definition of* β_1: Take $I \in \underline{\underline{\mathrm{Pic}}}_g(R)$; we define $\beta_1([I]) = [S \otimes I]$.

(5.5.4) *Definition of* γ_1: Let $I \in \underline{\underline{\mathrm{Pic}}}_g(S)$ represent an element of $H^0(S/R,\mathrm{Pic}_g)$. Then there exists a graded isomorphism φ :

$I_1 \to I_2$. Then $\varphi_2^{-1}\varphi_3\varphi_1$ is a graded isomorphism of I_{11}, so it is multiplication by a cocycle $t \in U(S^{(3)})$. It is easily verified that t is a cocycle, so we define $\gamma_1([I]) = [t]$.

(5.5.5) *Definition of* α_2: Let P be the S-module $S^{(2)}$, where S acts on the first factor. Then a cocycle $t \in U_0(S^{(3)})$ defines an $S^{(2)}$-module homomorphism $f(t) : P_1 \to P_2$, given by multiplication by t and switching the second and the third factor. Then $f(t)$ induces a graded descent datum $\varphi(t) : S \otimes \mathrm{END}_S(P) \to \mathrm{END}_S(P) \otimes S$, and defines a graded Azumaya algebra $A(t) = \{x \in \mathrm{END}_S(P) : \varphi(t)\varepsilon_1(x) = \varepsilon_2(x)\}$. Define $\alpha_2([t]) = [A(t)]$.

(5.5.6) *Definition of* β_2: For $[A] \in \mathrm{Br}_g(S/R)$, let $\sigma : S \otimes A \to \mathrm{END}_S(Q)$ be a graded splitting for A, with $\deg\sigma = 0$. Then $\varphi = \sigma_3\tau_1\sigma_2^{-1} : \mathrm{END}_{S^{(2)}}(Q_1) \to \mathrm{END}_{S^{(2)}}(Q_2)$ is induced by $f : Q_1 \otimes_2 I \to Q_2$, where $I \in \underline{\underline{\mathrm{Pic}}}_g(S \otimes S)$; I is a cocycle, and we define $\beta_2([A]) = [I]$.

(5.5.7) *Definition of* γ_2: Let I represent $[I] \in H^1(S/R, \mathrm{Pic}_g)$. Then there exists a graded $S^{(3)}$-isomorphism $f : I_1 \otimes_3 I_3 \to I_2$. We can check that $f_4^{-1}f_2^{-1}f_3f_1$ is just multiplication by a unit $u \in U_0(S^{(4)})$, which is a cocycle. Define $\gamma_2([I]) = [u]$. ■

III.5.6 Let $F : \underline{\underline{\mathrm{Pic}}}_g(R) \to \underline{\underline{\mathrm{Pic}}}^g(R)$ be the natural functor considered in III.3.4. Recall that there exists an isomorphism $\theta : \mathrm{gr}(R) \to K_1\Phi F$, given by $\theta(T) = [(R, Id, T)]$. The map $d : U(R) \to \mathrm{gr}(R) \cong K_1\Phi F$ maps u to $d(u) = R^u \cong [(R, Id, R^u)] = [(R, m(u^{-1}), R)]$, where $m(u^{-1})$ is just multiplication by u^{-1}. Also recall that for an isomorphism $f : P \to Q$ in R-mod, where P is a graded R-module, Q^f denotes the R-module Q with gradation induced by f, that is, $(Q^f)_i = f(P_i)$.

III.5.7 Proposition Let S be a graded faithfully flat R-algebra. Then

$$\mathrm{Im}(H^1_{\mathrm{gr}}(S/R, U) \longrightarrow H^1(S/R, U)) = \mathrm{Ker}(H^1(S/R, U)$$

$$\xrightarrow{d_1} H^1(S/R,\mathrm{gr})) = \mathrm{Pic}^g(S/R)$$

($\mathrm{Pic}^g(S/R)$ denotes $\mathrm{Ker}(\mathrm{Pic}^g(R) \to \mathrm{Pic}^g(S))$.)

Proof: The first equality follows from III.5.2: the sequence $H^1_{\mathrm{gr}}(S/R,U) \to H^1(S/R,U) \to H^1(S/R,\mathrm{gr})$ is exact. Consider $[t] \in \mathrm{Ker}(d_1)$; then there exists $T \in \mathrm{gr}(S)$ such that $d_1(t) = T_2 \otimes T_1^{-1}$, that is, $[(R,m(t^{-1}),R)] = [(R,Id,T_2 \otimes T_1^{-1})] = [(T_1,Id,T_2)]$, so $[(T_1,m(t),T_2)] = 1$ in $K_1\Phi F$, which means that $m(t) : T_1 \to T_2$ is graded. So $m(t)$ is a graded descent datum, as $t_2 = t_3t_1$. Hence $m(t)$ defines $I \in \underline{\underline{\mathrm{Pic}}}_g(R)$ such that $S \otimes I \cong_g T$. Put $\alpha_1([t]) = [I] \in \mathrm{Pic}^g(R) \subset \mathrm{Pic}(R)$. In fact, α_1 is the restriction of the well-known map $\alpha : H^1(S/R,U) \to \mathrm{Pic}(R)$ to $\mathrm{Ker}(d_1)$ (cf. [36, 37]), so α_1 is well defined, $\mathrm{Im}(\alpha_1) \subset \mathrm{Pic}^g(S/R)$, and α_1 is injective. Furthermore, $\mathrm{Im}(\alpha_1) = \mathrm{Pic}^g(S/R)$: let $I \in \underline{\underline{\mathrm{Pic}}}^g(R)$ represent $[I] \in \mathrm{Pic}^g(S/R)$; then there exists an isomorphism $f : S \otimes I \to S$. Let $T = S^f$; then $f : S \otimes I \to T$ is a graded isomorphism. Now define φ by commutativity of the following diagram of graded $S^{(2)}$-isomorphisms:

$$\begin{array}{ccc} I_{13} & \xrightarrow{f_1} & T_1 \\ \downarrow{\scriptstyle \tau_1} & & \downarrow{\scriptstyle \varphi} \\ I_{23} & \xrightarrow{f_2} & T_2 \end{array}$$

Then φ is multiplication by a unit $t \in U(S^{(2)})$, and t is a cocycle. Clearly $d_1(t) = [(T_1,Id,T_2)] = [(R,Id,T_2 \otimes T_1^{-1})]$, so $[t] \in \mathrm{Ker}(d_1)$. ■

II.5.8 Theorem (Second Graded Version of the Chase–Rosenberg Exact Sequence) Let R be a commutative graded ring, and S a commutative graded R-algebra which is faithfully projective as an R-module. Then we have an exact sequence

$$1 \longrightarrow H^0_{\mathrm{gr}}(S/R,U) \xrightarrow{\alpha_0} U(R) \xrightarrow{\beta_0} \mathrm{gr}(R)$$

$$\xrightarrow{\gamma_0} H^1_{\mathrm{gr}}(S/R,U) \xrightarrow{\alpha_1} \mathrm{Pic}^g(R) \xrightarrow{\beta_1} H^0(S/R,\mathrm{Pic}^g)$$

$$\xrightarrow{\gamma_1} H^2_{\mathrm{gr}}(S/R,U) \xrightarrow{\alpha_2} \mathrm{Br}_g(S/R) \xrightarrow{\beta_2} H^1(S/R,\mathrm{Pic}^g)$$

Proof: Since some parts of the proof involve complications of a graded nature, we prefer to present a detailed account of it. Note first that it follows from III.5.2 that we have an exact sequence of the form

$$1 \longrightarrow H^0_{\mathrm{gr}}(S/R,U) \xrightarrow{\alpha_0} H^0(S/R,U) \xrightarrow{\beta_0} H^0(S/R,\mathrm{gr})$$

$$\xrightarrow{\gamma_0} H^1_{\mathrm{gr}}(S/R,U) \longrightarrow H^1(S/R,U) \xrightarrow{d_1} H^1(S/R,\mathrm{gr}) \longrightarrow \cdots$$

It is well known that $H^0(S/R,U) = U(R)$, and from III.3.11, we retain that $H^0(S/R,\mathrm{gr}) \cong \mathrm{gr}(R)$. Using III.5.7, we obtain a map $\alpha_1 : H^1_{\mathrm{gr}}(S/R,U) \to \mathrm{Ker}(d_1) \cong \mathrm{Pic}^g(S/R) \subset \mathrm{Pic}(R)$, such that $\mathrm{Ker}(\alpha_1) = \mathrm{Im}(\gamma_0)$. The map $\beta_1 : \mathrm{Pic}^g(R) \to H^0(S/R,\mathrm{Pic}^g)$ is obtained by extension of scalars, so the exactness at $\mathrm{Pic}^g(R)$ follows immediately. This establishes the exactness of the sequence up to $\mathrm{Pic}^g(R)$.

(5.8.1) *Definition of* γ_1 *and exactness at* $H^0(S/R,\mathrm{Pic}^g)$: Let $I \in \underline{\underline{\mathrm{Pic}}}^g(S)$ represent $[I] \in H^0(S/R,\mathrm{Pic}^g)$. Let $\varphi : I_1 \to I_2$ be an $S^{(2)}$-isomorphism and $\mathrm{gr}(\varphi) = T$; then we get a graded isomorphism $\varphi : I_1 \to T \otimes_2 I_2$ (cf. III.3.5). The map $\varphi_2^{-1}\varphi_3\varphi_1$ is an isomorphism from I_{11} onto itself; hence it reduces to multiplication by a cocycle $t \in U(S^{(3)})$. Since $m(t) = \varphi_2^{-1}\varphi_3\varphi_1 : I_{11} \to D_1(T) \otimes_3 I_{11}$ is graded, we have that $[(S^{(2)}, Id, D_1(T))] = [(S^{(2)}, Id, D_1(T))][(I^*_{11}, m(t)^{-1}, I^*_{11} \otimes_3 D_1(T)^{-1})][(I_{11}, Id, I_{11})][(S^{(2)}, m(t), d_2(t))] = [(S^{(2)}, Id, d_2(t))]$ in $K_1\Phi F$, so $d_2(t) = D_1(T)$ in $\mathrm{gr}(S^{(2)})$. We define $\gamma_1([T]) = [(t,T)] \in H^2_{\mathrm{gr}}(S/R,U)$.

The morphism γ_1 is well defined by this. Indeed, let I' also represent [I], say $\varphi' : I'_1 \to I'_2$ [resp., $T' = \mathrm{gr}(\varphi')$] and $m(t') = \varphi_2'^{-1}\varphi'_3\varphi'_1$; then there exists an isomorphism $\psi : I \to I'$, and $\varphi'^{-1}\psi_2\varphi\psi_1^{-1} : I'_1 \to I'_1$ is an $S^{(2)}$-automorphism, so it is just multiplication by some $v \in U(S^{(2)})$. Computing yields $t' = (\Delta_1 v)t$

and $d_1(v) = \mathrm{gr}(\varphi'^{-1}\psi_2\varphi\psi_1^{-1}) = T'\otimes_2\mathrm{gr}(\psi)_2\otimes_2 T\otimes_2\mathrm{gr}(\psi)_1^{-1}$, so for $V = \mathrm{gr}(\psi)$, we have that $T'\otimes_2 T^{-1} = d_1(v)\otimes_2 D_0(V)$. Therefore $[(t,T)] = [(t',T')]$ indeed.

It is now easy to verify that $\gamma_1\beta_1 = 1$. On the other hand, assume that $\gamma_1([I]) = 1$ in $H^2_{\mathrm{gr}}(S/R,U)$; then, if (t,T) is constructed from I as above, we may write $t = \Delta_1 v$ and $T = d_1(v)\otimes_2 D_0(V^{-1})$. Consider $\varphi : I_1 \to T\otimes_2 I_2 = d_1 v\otimes_2 V_1^{-1}\otimes_2 V_2\otimes_2 I_2$; then $m(v^{-1}) : I_1\otimes_2 V_1 \to I_2\otimes_2 V_2$ is graded and it is a descent datum, inducing a graded R-module J with the property that $J\otimes S$ and $I\otimes_1 V$ are graded isomorphic. But then $[J\otimes S] = [I]$ in $\mathrm{Pic}^g(S)$, so $[I] = \beta_1([J])$, which proves the exactness at $H^0(S/R,\mathrm{Pic}^g)$.

(5.8.2) *Definition of* α_2 *and the exactness at* $H^2_{\mathrm{gr}}(S/R,U)$: Let $x\in H^2_{\mathrm{gr}}(S/R,U)$ be represented by (t,T); then multiplication by t induces graded morphisms $m(t) : S^{(3)} \to d_2(t) = D_1(T) = T_1\otimes_3 T_2^{-1}\otimes_3 T_3$ and $m(t) : T_1^{-1}\otimes_3 T_3^{-1} \to T_2^{-1}$. For $T\in\mathrm{gr}(S^{(n)})$, we let $P(T)$ be the graded S-module which is equal to T^{-1} as an abelian group, but with S acting only on the first factor. Although it is not an $S^{(3)}$-isomorphism, $\tau_1 m(t)$ induces a graded $S^{(2)}$-isomorphism $f(t) : P_1\otimes_2 T^{-1} \to P_2$, and $f(t)$ induces a graded $S^{(2)}$-isomorphism $\varphi(t) : \mathrm{END}_S(P_1) \to \mathrm{END}_S(P_2)$. It is easy to check that $f_2^{-1}f_3f_1$ is given by multiplying by $t_4\in U(S^{(4)})$ and since t_4 lies in the center of $\mathrm{END}_S(P_{11})$, it follows that $\varphi(t)$ is a graded descent datum for $\mathrm{END}_S(P)$, so it defines a graded Azumaya algebra $A(t)$. As $\theta : S\otimes A(t) \to \mathrm{End}_S(P) : s\otimes a \to sa$ is a graded isomorphism, we find that $[(A(t)]\in\mathrm{Br}_g(S/R)$. Moreover, $A(t)$ contains S as a maximal commutative graded subalgebra and $S\otimes 1$ is a maximal commutative graded subalgebra of $\mathrm{END}_S(P)$. We now define $\alpha_2(x) = [A(t)]$, and we claim that this yields a well-defined homomorphism $\alpha_2 : H^2_{\mathrm{gr}}(S/R,U) \to \mathrm{Br}_g(S/R)$. We shall verify this in several steps.

First, if $x = [(1,U_1\otimes U_2^{-1})]$, then (up to isomorphism) $A(1) = \mathrm{END}_R(Q)$ for some graded faithfully projective R-module Q. Indeed, we get $f(1) : P_1\otimes_2 U_1^{-1}\otimes_2 U_2 \to P_2$, which yields a graded descent datum $P_1 \otimes_2 U_1^{-1} \to P_2 \otimes_2 U_2^{-1}$. We thus obtain a graded faithfully projective R-module Q, such that the following diagram of graded $S^{(2)}$-module isomorphisms is commutative:

$$\begin{array}{ccc} Q_{13} & \longrightarrow & S \otimes (P \otimes_1 U^{-1}) \\ \downarrow & & \downarrow \\ Q_{23} & \longrightarrow & (P \otimes_1 U^{-1}) \otimes S \end{array}$$

Now, the identity on $P \otimes_1 U^{-1}$ induces a graded $S^{(2)}$-isomorphism $\psi : \mathrm{END}_S(P) \to \mathrm{END}_S(P \otimes_1 U^{-1})$ (which is again just the identity, if we neglect gradations) and we obtain a commutative diagram

$$\begin{array}{ccccc} S \otimes \mathrm{END}_R(Q) \otimes S & \longrightarrow & S \otimes \mathrm{END}_S(P \otimes_1 U^{-1}) & \longrightarrow & S \otimes \mathrm{END}_S(P) \\ \downarrow & & \downarrow & & \downarrow \\ \mathrm{END}_R(Q) \otimes S \otimes S & \longrightarrow & \mathrm{END}_S(P \otimes_1 U^{-1}) \otimes S & \longrightarrow & \mathrm{END}_S(P) \otimes S \end{array}$$

It follows that $\mathrm{END}_R(Q)$ is a descended algebra for $\mathrm{END}_S(P)$ and a uniqueness argument yields that $\mathrm{END}_R(Q)$ and $A(t)$ are graded isomorphic. Next, let $x = [(t,T)]$ and $x' = [(t\Delta_1 s, T \otimes_2 d_1 s)] \in H^2_{\mathrm{gr}}(S/R,U)$, where $s \in U(S^{(2)})$; then $m(s)A(t\Delta_1 s)m(s^{-1}) = A(t)$. Indeed, $z \in \mathrm{END}_S(P)$ lies in $m(s^{-1})A(t)m(s)$ if and only if $m(s)zm(s^{-1}) \in A(t)$, that is, $\tau_1 t s_1 z s_1^{-1} \tau_1 = s_3 z_2 s_3^{-1}$. Using the fact that s_3 lies in the center of $\mathrm{END}_{S^{(2)}}(P_1)$ and $\tau_1 s_2 = s_3$, we obtain that this is equivalent to $z_2 = \tau_1 t s_1 s_2^{-1} z_1 s_2 s_1^{-1} t^{-1} \tau_1 = \tau_1 t s_1 s_2^{-1} s_3 z_1 s_3^{-1} s_2 s_1^{-1} t^{-1} \tau^{-1}$ or $z \in A(t\Delta_2 s)$. Moreover, $m(s) : T \to T \otimes_2 d_1(s)$ is a graded isomorphism; hence, $m(s^{-1}) : P' = P(T \otimes_2 d_1(s)) \to P = P(T)$ is graded, inducing a graded isomorphism $\mathrm{END}_S(P') \to \mathrm{END}_S(P)$, which sends z to $m(s^{-1})zm(s)$. The restriction to $A(t') \otimes A(t)$ yields an R-algebra isomorphism of degree zero.

Now, let $x = [(t,T)]$ and $x^{-1} = [(t^{-1},T^{-1})] \in H^2_{\text{gr}}(S/R,U)$; then we claim that $A(t) \otimes A(t^{-1})$ is graded isomorphic to $\text{END}_R(Q)$ for some faithfully projective graded R-module Q. Indeed, the graded isomorphism $g = f(t) \otimes f(t^{-1}) : P(T)_1 \otimes_2 P(T^{-1})_1 \otimes_2 T^{-1} \otimes_2 T \to P(T)_2 \otimes P(T^{-1})_2$ induces an isomorphism $\text{END}_{S^{(2)}}(P(T)_1 \otimes_2 P(T^{-1})_1) \to \text{END}_{S^{(2)}}(P(T)_2 \otimes_2 P(T^{-1})_2)$, which is a graded descent datum of degree zero, defining $A(t) \otimes A(t^{-1})$. Note that $P(T) \otimes_1 P(T^{-1})$ is graded isomorphic to $P(T_3 \otimes_3 T_2^{-1})$, and using the fact that $f(t)_2^{-1} f(t)_3 f(t)_1$ is given by multiplication by t_4, we easily deduce that $g_2^{-1} g_3 g_1$ is multiplication by $t_{45} t_{45}^{-1} = 1$. So g is a graded descent datum for $P(T_3 \otimes_3 T_2^{-1})$, defining a graded R-module Q such that the following diagram of graded $S^{(2)}$-isomorphisms commutes:

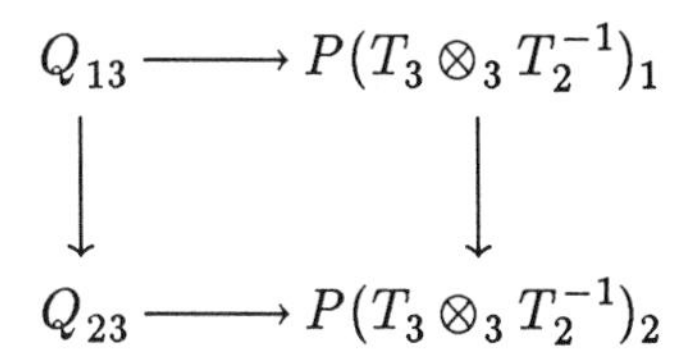

which induces

$$\begin{array}{ccc} \text{END}_R(Q)_{13} & \longrightarrow & \text{END}_R(P(T_3 \otimes_3 T_2^{-1})_1) \\ \downarrow & & \downarrow \\ \text{END}_R(Q)_{23} & \longrightarrow & \text{END}_R(P(T_3 \otimes_3 T_2^{-1})_2). \end{array}$$

Comparing this diagram with the commutative diagram which defines $A(t) \otimes A(t^{-1})$ yields that $A(t) \otimes A(t^{-1})$ is graded isomorphic to $\text{END}_R(Q)$, by a uniqueness argument. Finally, if $y = [(t,T)]$ and $y' = [(t',T')]$ are in $H^2_{\text{gr}}(S/R,U)$, then a similar argument shows that $A(t) \otimes A(t') \otimes A((tt')^{-1})$ is of the form $\text{END}_R(Q)$ for some faithfully projective graded R-module Q, proving the assertion.

Now, it is easy to see that $\alpha_2 \circ \gamma_1 = 1$, so in order to prove exactness at $H^2_{\text{gr}}(S/R,U)$, we have to show that $\text{Ker}(\alpha_2) \subset \text{Im}(\gamma_1)$.

Suppose that $\alpha_2(x) = 1$ for some $x = [(t,T)] \in H^2_{\mathrm{gr}}(S/R,U)$, and let $\rho : A(t) \to \mathrm{END}_R(Q)$ be some graded isomorphism for some $Q \in \underline{\underline{FP}}^g(R)$. Consider $\zeta : S \otimes A(t) \to \mathrm{END}_S(P)$, where $P = P(T)$ and $\psi = (1 \otimes \rho) \circ \zeta^{-1} : \mathrm{END}_S(P(T)) \to \mathrm{END}_S(S \otimes Q)$. This map is induced by $\lambda : P \otimes_1 I \to S \otimes Q$ for some $I \in \underline{\underline{\mathrm{Pic}}}^g(S)$. Let $g = h_2\tau_1 h_1^{-1} : P_1 \otimes_2 I_1 = S \otimes (P \otimes_1 I) \to P_2 \otimes_2 I_2 = (P \otimes_1 I) \otimes S$, which is clearly a descent datum for $P \otimes_1 I$ and thus defines Q. One verifies easily that it induces the map $\psi_3^{-1}\tau_1\psi_1 = \zeta_3\tau_{13}\tau_1\tau_{11}^{-1}\zeta_1 = \zeta_3\tau_1\zeta_1^{-1} = \varphi(t)$, hence $g : P_1 \otimes_2 I_1 \to P_2 \otimes_2 I_2$ and $\varphi(t) : P_1 \otimes_2 T^{-1} \to P_2$ induce the same map $\varphi(t)$. Thus we find a graded isomorphism $\varphi : I_1 \otimes_2 I_2^{-1} \to T^{-1}$ such that $g = f(t) \otimes_2 \varphi$. If one writes down the descent condition for g, then one obtains that $t_1^{-1}t_3^{-1}t_2\varphi_2^{-1}\varphi_3\varphi_1$ is the identity on $(P \otimes I)_{11}$ and since $t_1 t_2^{-1} t_3$ is multiplication by t_4 and $\varphi_2^{-1}\varphi_3\varphi_1$ is multiplication by t, it follows that $x = \gamma_1([I^{-1}])$.

(5.8.3) *Definition of* β_2 *and exactness at* $\mathrm{Br}_g(S/R)$: Let A represent an element $[A] \in \mathrm{Br}_g(S/R)$ and let $\sigma : S \otimes_R A \to \mathrm{END}_S(Q)$ be a graded splitting for A. Write $\varphi = \sigma_3\sigma_1\sigma_2^{-1} : \mathrm{END}_{S^{(2)}}(Q_1) \to \mathrm{END}_{S^{(2)}}(Q_2)$; then φ is graded and induced by a graded $S^{(2)}$-isomorphism $f : Q_1 \otimes_2 I \to Q_2$ for some $I \in \underline{\underline{\mathrm{Pic}}}^g(S^{(2)})$. Since $\varphi_2 = \varphi_3\varphi_1$, we find that $[I_2] = [I_3 \otimes_3 I_1]$ in $\mathrm{Pic}^g(S^{(3)})$; hence we may write $\beta_2([A]) = [I] \in H^1(S/R,\mathrm{Pic}^g)$. An easy verification yields that β_2 is well defined and that $\beta_2 \circ \alpha_2 = 1$.

On the other hand, if $\beta_2([A]) = [I] = 1$ in $H^1(S/R,\mathrm{Pic}^g)$, with I constructed as above, then there exists $J \in \underline{\underline{\mathrm{Pic}}}^g(S)$ and an $S^{(2)}$-isomorphism $\psi : I \to J_1 \otimes_2 J_2^{-1}$. If $T = \mathrm{gr}(\psi)$, then we have a graded isomorphism $\psi : I \to J_1 \otimes_2 J_2^{-1} \otimes_2 T$. Now, ψ may be induced by a graded isomorphism $g : (Q \otimes_1 J)_1 \otimes_2 T \to (Q \otimes_1 J)_2$. Put $U = Q \otimes_1 J$ and consider the graded map $g_2^{-1}g_3g_1 : U_{11} \otimes_3 T_1 \otimes_3 T_3 \otimes_3 T_2^{-1} \to U_{21} \otimes_3 T_3 \otimes_3 T_2^{-1} = U_{13} \otimes_3 T_3 \otimes_3 T_2^{-1} \to U_{23} \otimes_3 T_2^{-1} \to U_{11}$. It is just multiplication by some unit $t \in U(S^{(3)})$, which is a cocycle, and we have

$d_2(t) = D_1(T)$, so (t,T) represents an element of $H^2(S/R,U)$. Let $h(t^{-1})$ be the canonical splitting map for $A(t^{-1})$ and $P = P(T^{-1})$. Consider $\sigma \otimes_1 h(t^{-1}) : S \otimes A \otimes A(t^{-1}) \to \mathrm{END}_S(Q \otimes_1 P)$ and note that $\varphi \otimes_1 f(t^{-1})$ is induced by $h = g \otimes_2 f(t^{-1}) : (U \otimes_1 P)_1 \to (U \otimes_1 P)_2$. Since $h_2^{-1} h_3 h_1$ is multiplication by $tt^{-1} = 1$, h is a graded descent datum for $U \otimes_1 P$ with descended module M, say. A uniqueness argument yields that $A \otimes A(t^{-1})$ and $\mathrm{END}_R(M)$ are graded isomorphic, so $[A] = [A(t^{-1})]$ in $\mathrm{Br}_g(S/R)$. ■

III.5.9 Corollary (Crossed Product Theorems) Let S be a commutative graded R-algebra, and suppose that S is faithfully projective as an R-module.

(5.9.1) If $\mathrm{Pic}_g(S) = \mathrm{Pic}_g(S \otimes S) = 1$, then $\mathrm{Br}_g(S/R) \cong H^2(S/R,U_0)$.

(5.9.2) If $\mathrm{Pic}^g(S) = \mathrm{Pic}^g(S \otimes S) = 1$, then $\mathrm{Br}_g(S/R) \cong H^2_{\mathrm{gr}}(S/R,U)$. ■

III.5.10 Note that if S is a graded Galois extension of R with Galois group $G = \mathrm{Gal}(S/R)$, then the cohomology groups used previously may be replaced by Galois cohomology groups. Let us describe the map $\alpha_2 : H^2_{\mathrm{gr}}(G,U(S)) \to \mathrm{Br}_g(S/R)$ in this case. For $\sigma \in G$ and $I \in \underline{\underline{\mathrm{Pic}}}^g(S)$, we let ${}_\sigma I$ be the usual graded S-module obtained from I by twisting by σ, that is, $s \cdot x = \sigma^{-1}(s)x$ for all $s \in S$, $x \in I$. This yields an action of G on $K_1\Phi F(S)$ (and on $\mathrm{gr}(S)$) by putting $\sigma \cdot [(P,f,Q)] = [({}_\sigma P, f, {}_\sigma Q)]$. It is clear that $K_1\Phi F(S)$ and $\mathrm{gr}(S)$ are invariant under the action of G. Now, an element of $H^2_{\mathrm{gr}}(G,U(S))$ may be represented by a couple (c,F), where $c : G \times G \to U(S)$ and $F : G \to \mathrm{gr}(S)$ have the following properties:

(i) $c(\rho\sigma,\tau)c(\rho,\sigma) = c(\rho,\sigma\tau)\rho \cdot c(\sigma,\tau)$ for all $\rho,\sigma,\tau \in G$.

(ii) $d(c(\sigma,\tau)) \otimes_1 F(\sigma) \otimes_1 F(\tau) = F(\sigma\tau)$ for all $\sigma,\tau \in G$.

To (c,F) we associate the Azumaya algebra $A = A(c,F) = \oplus_{\sigma \in G} F(\sigma)$, with multiplications $s_\sigma u_\sigma s_\tau u_\tau = s_\sigma \sigma(s_\tau) c(\sigma,\tau) u_{\sigma\tau}$ for any $\sigma,\tau \in G$ and $s_\sigma \in F(\sigma)$ [resp., $s_\tau \in F(\tau)$]—the u_σ are the canonical basic vectors. The gradation on A is given by $A_i = \oplus_{\sigma \in G} F(\sigma)_i$. The fact that A is an associative algebra follows from (i). That it is graded follows from (ii): Consider $s_\sigma u_\sigma \in F(\sigma)_i$, $s_\tau u_\tau \in F(\tau)_j$; then $s_\sigma \sigma(s_\tau) f(\sigma,\tau) \in F(\sigma\tau)_{i+j}$, because of (ii) and the fact that $\deg \sigma = 0$.

III.5.11 Let us now look at the case where R is reduced and S does not contain nontrivial idempotents; then we claim that S is also reduced. Indeed, the trace homomorphism tr is a free generator for the right S-module $\mathrm{Hom}_R(S,R) = S^*$, and since S is a Galois extension of R, we have that $\mathrm{tr}(a) = \sum \sigma(a)$, where σ runs through G, for all a in S. If a is nilpotent, then for all x in S we get that $\mathrm{tr}(ax) = \sum \sigma(ax)$ is nilpotent, hence $\mathrm{tr}(ax) = 0$. Since tr is a free generator for S^*, it follows that $a = 0$ indeed.

So, let R be reduced and assume that S has no nontrivial idempotents; then S is reduced, $\mathrm{gr}(S) = \mathbf{Z}$, and all invertible elements of S are homogeneous. The map $d : U(S) \to \mathrm{gr}(S) = \mathbf{Z}$ is defined by $d(u) = -\deg u$, for $\deg_{d(u)}(u) = 0$, hence $\deg_{d(u)} 1 = \deg_R u^{-1} + \deg_{d(u)} u = -\deg_R u$. Moreover, $H^1(G,\mathrm{gr}(S)) = H^1(G,\mathbf{Z}) = 1$ as G is a torsion group and $\mathbf{Z}$ is torsion free, so we have an injection of $H^2_{\mathrm{gr}}(G,U(S))$ into $H^2(G,U(S))$; hence $H^2_{\mathrm{gr}}(G,U(S)) = \mathrm{Ker}(H^2(G,U(S)) \to H^2(G,\mathbf{Z}))$.

III.5.12 Corollary Let R be a reduced graded ring, and S a graded Galois extension of R, which has no nontrivial idempotents. Then $\mathrm{Pic}^g(S) = 1$ implies that $\mathrm{Br}_g(S/R) \cong \mathrm{Ker}(H^2(G,U(S)) \to H^2(G,\mathbf{Z}))$.

III.5.13 Example Let $R = \mathbf{R}[T,T^{-1}]$ with $\deg T = 1$. Consider the graded Galois extension $S = \mathbf{C}[T,T^{-1}]$ of R with

$\mathrm{Gal}(S/R) = \{\sigma_0, \sigma_1\} = \{1, ^-\}$. Clearly the graded Azumaya algebra $\mathbf{H}[T, T^{-1}]$ is a crossed product over the cocycle $f(\sigma_0, \sigma_i) = 1$, $f(\sigma_1, \sigma_1) = -1$. Indeed, $\mathrm{Pic}_g(S) = 1$, so every element of $\mathrm{Br}_g(S/R)$ may be written as a crossed product over elements of degree zero.

Let R', S' be the same rings as above, but with $\deg T = 2$; then $B = \mathbf{C}[X, X^{-1}, ^-, X^2 = T]$ is a graded Azumaya algebra represented by the cocycle $g(\sigma_0, \sigma_i) = 1, g(\sigma_1, \sigma_1) = T$, which is not of degree zero, but which lies in $\mathrm{Ker}(d)$. Indeed, $\mathrm{Pic}_g(S') = \mathbf{Z}/2\mathbf{Z}$. Yet $\mathrm{Pic}^g(S') = 1$, so we may apply III.5.12, that is, $\mathrm{Br}_g(S'/R') = H^2_{\mathrm{gr}}(G, U(S')) = \{1, [A], [B], [A \otimes B]\}$. However, B can be made equivalent to a crossed product of degree zero: Let $S'' = \mathbf{C}[Y, Y^{-1}, Y^2 = T]$; then $\mathrm{Pic}_g(S'') = 1$ and B is equivalent to the crossed product over the Galois cocycle f given by $f(\sigma_1, \sigma_2) = f(\sigma_2, \sigma_1) = f(\sigma_2, \sigma_2) = -1$, and $f(\sigma_i, \sigma_j) = 1$ for other combinations.

Chapter IV

Application to Some Special Cases

IV.1 BRAUER GROUPS OF GRADED FIELDS

IV.1.1 All gradations considered in this chapter are $\mathbf{Z}$-gradations. Recall that in this case a (commutative) graded field is of the form k or $k[T,T^{-1}]$, where k is a field and T a variable of degree $t \in \mathbf{N}_0$. An arbitrary graded field is of the form D or $D[X,X^{-1},\varphi]$ where D is a skewfield, φ an automorphism of D, and X a variable of positive degree satisfying the commutation rule $Xa = a^{\varphi}X$ for any $a \in D$. Let R be a graded ring and $\underline{d} \in \mathbf{Z}^n$. As before, we denote by $R^n(\underline{d})$ the graded free R-module with homogeneous generators $e_1, e_2, \ldots, e_n$, where $\deg e_i = d_i$. We let $M_n(R)(\underline{d}) = \mathrm{END}_R(R^n(\underline{d}))$. Then $M_n(R)(\underline{d})$ is in fact the matrix ring $M_n(R)$ with gradation given by $M_n(R)(\underline{d})_\alpha = \{A \in M_n(R) : \deg A_{ij} = \alpha + d_j - d_i\}$.

A graded ring S is called *gr-simple* if it contains no nontrivial two-sided graded ideal. It is a well-known result that there exists a graded counterpart to the classical Wedderburn–Artin theorem: Every gr-simple gr-artinian ring is graded isomorphic to $M_n(\Delta)(\underline{d})$ for some graded (skew)field Δ. For the proof, we refer to [48, II.9]. For a commutative graded field K, we call A a *gr-central simple* K-algebra if A is gr-simple and finite over K, and $Z(A) = K$.

IV.1.2 Proposition Let K be a graded field and A a gr-central simple K-algebra. Then A_0 is a central $Z(A_0)$-simple algebra if it is a prime ring.

Proof: Of course, we may suppose that K is nontrivially graded, so $K = k[T,T^{-1}]$, $A = M_n(\Delta)$, $\Delta = D[X,X^{-1},\varphi]$. Since A is a PI-algebra, D is finite dimensional over $k = Z(D)^{\varphi}$. Now $(A_0)_{ij} = DX^{d_i - d_j}$, hence A_0 is finite dimensional over k, that is, artinian. ∎

In the foregoing proposition we may replace the condition that A_0 is a prime ring by the condition that A is uniformly gr-simple in the sense of [48]. In fact these conditions are equivalent and also equivalent to the condition that for every n such that $A_n \neq 0$ we also have that $\Delta_n \neq 0$, or to the condition that A is a direct sum of graded minimal left ideals that are isomorphic in A-gr. All of this may be formulated in an Artin–Wedderburn theorem for small additive categories.

IV.1.3 Proposition If K is a commutative graded field, then A is a graded K-Azumaya algebra if and only if A is a gr-central simple K-algebra.

Proof: Suppose that A is a graded K-Azumaya algebra; then by [23, II.3.7], there exists a one-one correspondence between ideals of K and two-sided ideals of A, given by $a \to aA$, and $I \to I \cap R$. Since K has no graded ideals, A is therefore gr-simple. Since

A is a faithfully projective K-module ([37, II.5.1]), A is finite dimensional over K; hence A is gr-artinian.

Conversely, let A be a gr-central simple K-algebra. Then a slight adaptation of the proof of [29, IV.1.3] yields that $A \otimes_K A^{\mathrm{opp}} \cong_g \mathrm{END}_R(A)$, where the isomorphism is given by $a \otimes b \to axb$. The result then follows from [37, III.5.1]. ■

IV.1.4 Proposition For every $\underline{d} \in Z^n$ and if $\deg X = 1$, we have that $M_n(D[X,X^{-1},\varphi])(\underline{d})_0 \cong M_n(D)$, that is, the different gradations possible on $M_n(D[X,X^{-1},\varphi])$ which extend the gradations on the center $k[T,T^{-1}]$ determine isomorphic parts of degree zero.

Proof: Let us write $e_{ij} = d_i - d_j$, for $i,j \in \{1,\ldots,n\}$. Define a map $\pi : (M_n(D[X,X^{-1},\varphi])(\underline{d})_0 \to M_n(D)$ by sending the matrix with entries $\{a_{ij}X^{e_{ij}} : 1 \leq \cdots \leq n\}$ to the matrix with entries $\{\varphi^{e_{1j}}(a_{ij}) : 1 \leq \cdots \leq n\}$. Obviously, π is bijective, and, using the fact that $e_{ij} + e_{jk} = e_{ik}$, we obtain $\pi((a_{ij}X^{e_{ij}})(b_{k1}X^{e_{k1}})) = \sum_k \varphi^{e_{1i}}(a_{ik})\varphi^{e_{1k}}(b_{kj})$. On the other hand $\pi((a_{ij}X^{e_{ij}}))\pi((b_{k1}X^{e_{k1}})) = (\varphi^{e_{1i}}(a_{ij}))\varphi^{e_{1k}}(b_{kj})) = \sum_k \varphi^{e_{1i}}(a_{ik})\varphi^{e_{1k}}(b_{kj})$. Hence π is a ring isomorphism. ■

IV.1.5 Corollary It is easily verified that, for every $\underline{a} \in M_n(\Delta)(\underline{d})_0 = A_0$, $\pi(\underline{a}) = \underline{xax}^{-1}$, where $\underline{x}$ is the diagonal matrix with entries $x_{ii} = X^{e_{1i}}$. The automorphism $\varphi : D \to D$ may be extended to an automorphism $\tilde{\varphi} : A_0 \to A_0$ and we obtain that $A = A_0[X,X^{-1},\tilde{\varphi}]$. On the other hand, φ extends to an automorphism of $M_n(D)$, $\hat{\varphi}$ say, and also $A = M_n(D)[\underline{y},\underline{y}^{-1},\hat{\varphi}]$, where $\underline{y} = X\underline{x}$. It is clear that $\hat{\varphi}$ is nothing but conjugation by $\underline{y}$ composed with $\tilde{\varphi}$. ■

Before we prove the splitting theorem for gr-central simple algebras, let us first recall the following:

IV.1.6 Proposition Let S be a maximal commutative subalgebra of a graded R-Azumaya algebra A, and suppose that S is graded as an R-algebra. Then we have a graded isomorphism $A \otimes_R S \cong_g \text{END}_S(A)$, where A is a graded S-module by multiplication on the right. If A is projective as an R-module, then S is projective as an R-module, and in this case $[A : R] = [A : S]^2$ and $[S : R] = [A : S]$. If S is separable as an R-algebra, then A is projective as an S-module.

Proof: This is II.6.1 in [37], up to the trivial observation that the isomorphism $A \otimes_R S \cong \text{END}_S(A)$ is graded. ■

IV.1.7 Theorem Let $K = k[T,T^{-1}]$ be a graded field. For every gr-central simple K-algebra A, there exists a maximal commutative subalgebra of A, which is graded and separable and of the form $k(\alpha)[T,T^{-1}]$, where $\deg \alpha = 0$. Consequently, every gr-central simple K-algebra may be split in $\text{Br}_g(K)$ by a gr-Galois extension of the form $l[T,T^{-1}]$.

Proof: Let us first assume that $A = \Delta = D[X,X^{-1},\varphi]$ is a graded skewfield. Suppose that $\deg T/\deg X = n$, and write $Z(D) = k_1$. Then $k_1^{\varphi} = k$, and k_1 is a Galois extension of k with Galois group $\{1,\varphi,\ldots,\varphi^{m-1}\}$, where m is the least positive integer for which $\varphi^m \mid_{k_1} = 1$. This may be checked easily using the last condition in [23, III.1.2]. Let l be a maximal commutative subfield of D, and suppose that l/k_1 is separable. Then, of course, l splits D in $\text{Br}(k_1)$, and $[l : k_1] = [D : l] = p$. Write $L = l[T,T^{-1}]$, $K_1 = k_1[T,T^{-1}]$; then $[\Delta : L] = [\Delta : D[T,T^{-1}]][D[T,T^{-1}] : l[T,T^{-1}]] = np$, and $[L : K] = [L : K_1][K_1 : K] = pm$. Clearly $m \leq n$.

Consider the graded skewfields $B = D[X^m,X^{-m},\varphi^m] \subset C = D[T,T^{-1}]$. As $\varphi^m = 1$ on k_1, it follows that $Z(B) = Z(C) = K_1$, so B and C are gr-central simple K_1-algebras. As $C^B = K_1$, then $B = C^{K_1} = C$ by the double commutator theorem, and $m = n$.

It now follows from (IV.1.6) that L is a maximal commutative subalgebra of Δ, so we have shown the first assertion in the case where A is a graded skewfield. The second assertion now follows if we embed l into a Galois extension of k.

We still have to prove the first statement in the case where $A = M_r(\Delta)$. Suppose that k is infinite. Choose $\beta \in D$ such that $k(\beta)$ is a maximal commutative subfield of D which is a separable extension of k. The let α be a diagonal matrix of A having r different entries in $k(\beta)$. Then $K[\alpha]$ is the graded subalgebra of A consisting of all diagonal elements with entries in $k(\beta)[T,T^{-1}]$; hence it is a maximal commutative subalgebra of A, and a graded separable extension of K. Finally, suppose that k is finite. Then $\Delta = l[X,X^{-1},\varphi]$ for some separable extension l of k (l/k is even Galois, since $l^{\varphi} = k$). Let l' be a field extension of l of degree r. L acts on Δ by right multiplication, so $L' = l'[T,T^{-1}]$ defines an action on Δ^r. So L' may be identified with a subalgebra of $M_r(\Delta)$. By a rank argument, it is a maximal one. ■

IV.1.8 Theorem Let $K = k[T,T^{-1}]$, with $\deg T = t$, be a graded field. Then every graded Azumaya algebra over K may be represented in $\mathrm{Br}_g(K)$ by a Galois crossed product over a Galois extension L of the form $l[T,T^{-1}]$. The generating elements may be chosen such that they are homogeneous. Moreover,

$$\begin{aligned}\mathrm{Br}_g(K) &= H^2_{\mathrm{gr}}(\mathrm{Gal}(k^{\mathrm{sep}}/k), U(k^{\mathrm{sep}}[T,T^{-1}])) \\ &= \mathrm{Br}(k) \oplus H^2_{\mathrm{gr}}(\mathrm{Gal}(k^{\mathrm{sep}}/k), t\mathbf{Z}).\end{aligned}$$

If $\mathrm{char}(k)$ does not divide $\deg T = t$, that is, if K is quasistrongly graded, then every graded Azumaya algebra may be represented by a crossed product over elements of degree zero where the generating elements are also of degree zero. We then have $\mathrm{Br}_g(K) = \varinjlim H^2(\mathrm{Gal}(L/K), U(L_0)) = \mathrm{Br}(k) \oplus H^2(\mathbf{Z}/t\mathbf{Z}, (k^{\mathrm{sep}})^*)$, where the limit is taken over all graded Galois extensions L of K.

Proof: From the foregoing theorem, we retain that every K-Azumaya algebra has a graded Galois splitting ring of the form $L = l[T,T^{-1}]$. Let $G = \text{Gal}(l/k)$. As $\text{Pic}^g(L) = 1$, $\text{Br}_g(L/K) = H^2_{\text{gr}}(G,U(L))$, so A is gr-equivalent to a crossed product of the desired form, by (III.5.12). Now $H^2_{\text{gr}}(G,U(L)) = H^2_{\text{gr}}(G,l^* \times \langle T \rangle) = H^2_{\text{gr}}(G,l^*) \oplus H^2_{\text{gr}}(G,\langle T \rangle)$. Identifying T with $t\mathbf{Z}$ by the degree map, we obtain the result after taking limits over all Galois extensions l of k. Note that $H^2_{\text{gr}}(G,\mathbf{Z})$ is in fact the kernel of the map $H^2(G,t\mathbf{Z}) \to H^2(G,\mathbf{Z})$.

Next, suppose that $\text{char}(K)$ does not divide t. Let A be split by $l[T,T^{-1}]$, and replace l by a cyclotomic extension containing all tth roots of 1. Then $L' = L/(X^t - T)$ is a gr-Galois extension of L with Galois group $G = \{1,\tau_1,\ldots,\tau_{t-1}\}$, where τ_i is determined by $\tau_i(X) = \delta^i X$, where 1, δ, δ^2, ... are the tth roots of 1 in l. This may be checked using condition (iii) in [23, III.1.3]. As $\text{Pic}_g(L') = 1$, $\text{Br}_g(L'/K) = H^2(\text{Gal}(L'/K),l^*) = H^2(\text{Gal}(l/k),l^*) \oplus H^2(G,l^*)$. Then take limits over all graded Galois extensions of K. ∎

IV.1.9 Proposition Let $K = k[T,T^{-1}]$ be a graded field, where $\deg T = t$. Then every graded Azumaya algebra is equivalent in $\text{Br}_g(k[T,T^{-1}])$ to a Galois crossed product A over a Galois extension $L = l[T,T^{-1}]$ of K, where the generating elements $\{u_\sigma : \sigma \in G\}$ and the cocycle elements $c_{\sigma\tau}$ are homogeneous. If $G_0 = \{\sigma : \deg u_\sigma = 0 \bmod t\}$, then G_0 is a normal subgroup of G, and G/G_0 is t-torsion. If $(t,|G|) = 1$, then $A = A_0[T,T^{-1}]$.

Proof: The first statement follows from the theorem and (III.5.11). We remark that $\deg c_{\sigma\tau} \in t\mathbf{Z}$, since $c_{\sigma\tau} \in L$. It follows that $\deg u_\sigma + \deg u_{\sigma^{-1}} = \deg c_{\sigma,\sigma^{-1}} + \deg u_1 = 0 \bmod t$ (we may always choose $u_1 = 1$). So, for $\sigma \in G_0$, $\tau \in G$, $\deg(u_\tau u_\sigma u_{\tau^{-1}}) = \deg c_{\tau\sigma} + \deg c_{\tau\sigma,\tau^{-1}} + \deg u_{\tau\sigma\tau^{-1}}$, and it follows that $\deg u_{\tau\sigma\tau^{-1}} = 0 \bmod t$, so $\tau\sigma\tau^{-1} \in G_0$. Also, for

$\sigma \in G$, $u_\sigma^t = c_{\sigma,\sigma} c_{\sigma^2,\sigma} c_{\sigma^{t-1},\sigma} u_{\sigma^t}$, so $\deg u_{\sigma^t} = 0 \bmod t$, and $\sigma^t \in G_0$.

Finally, let $g = |G|$, $g_0 = |G_0|$. Then $(t,g) = (t,g_0) = 1$, and every $\sigma \in G_0$ has a tth root in G_0. Take $\tau \in G$; then $\tau^t \in G_0$, and $\tau^t = \gamma^t$ for some $\gamma \in G_0$. Actually γ may taken to be $(\tau^t)^\alpha$, where $\alpha t + \beta g = 1$. Therefore τ and γ commute and $\tau^t = \gamma^t$ entails $\tau = \gamma$, so $\tau \in G_0$, and $G = G_0$. The result follows. ∎

IV.1.10 Example The second part of the theorem does not hold if $\mathrm{char}(k) \mid \deg T$. Let $k = \mathbf{F}_2$, $K = k[T,T^{-1}]$, where $\deg T = 2$. Let $l = k(\alpha)$, where α is a third root of unity, so $\alpha^2 + \alpha + 1 = 0$. Then $L = l[T,T^{-1}]$ is a graded Galois extension of K with group $G = \{1,\sigma\}$ where σ is determined by $\sigma(\alpha) = \beta = 1 + \alpha = 1/\alpha$.

The graded K-Azumaya algebra $A = l[X,X^{-1},X^2 = T,\sigma]$ is written as a crossed product over the cocycle $f(\sigma,\sigma) = T$, $f(1,\sigma) = f(\sigma,1) = f(1,1) = 1$. It cannot be written as a crossed product in degree zero, since every graded separable extension of K is of the form $l[T,T^{-1}]$ (cf. ch. II).

For a graded ring R, and for an integer n, recall that $R_{(n)}$ is the graded ring determined by $(R_{(n)})_{ni} = R_i$ (blow up the degree in R by a factor n). Obviously, if m and n are positive integers, then we have canonical maps $\mathrm{Br}_g(R_{(m)}) \to \mathrm{Br}_g(R_{(mn)})$. We define $U\mathrm{Br}_g(R) = \varinjlim \mathrm{Br}_g(R_{(n)})$.

IV.1.11 Theorem Suppose that k is a perfect field. Then $\mathrm{Br}(k[T,T^{-1}]) = U\mathrm{Br}_g(k[T,T^{-1}])$.

Proof: It is well known that the map $\mathrm{Br}(k[T,T^{-1}]) \to \mathrm{Br}(k(T))$ is injective; indeed, $k[T,T^{-1}]$ is a regular ring. Also, by Tsen's theorem, every $k(T)$-Azumaya algebra may be split by a Galois extension of the form $l[T,T^{-1}] = L$. Since $\mathrm{Pic}(L) = \mathrm{Pic}^g(L) = 1$, $\mathrm{Br}(L/K) = H^2(G,U(L))$, and $\mathrm{Br}_g(L/K) = H^2_{\mathrm{gr}}(G,U(L_{(n)}) =$

$\mathrm{Ker}(H^2(G,U(L_{(n)}) \to H^2(G,\mathbf{Z}))$, by II.5.11. Let $n = |G|$; then the image of a cocyle in $Z^2(G,U(L_{(n)})$ takes values only in $n\mathbf{Z}$. But in $H^2(G,\mathbf{Z})$ every element has order $|G|$, so a cocycle taking values in $n\mathbf{Z}$ represents the trivial element of $H^2(G,\mathbf{Z})$. Consequently $H^2_{\mathrm{gr}}(G,U(L_{(n)}) = H^2(G,U(L))$, and $\mathrm{Br}_g(L_{(n)}/K_{(n)}) = \mathrm{Br}(L/K)$. The result follows if we take the limit over all gr-Galois extensions L. ■

IV.1.12 Proposition Let k be a nonperfect field. Then $U\mathrm{Br}_g(k[T,T^{-1}]) \neq \mathrm{Br}(k[T,T^{-1}])$.

Proof: We have to construct a K-Azumaya algebra which cannot be graded, even if we blow up the degree of T. Such an algebra was given by Auslander and Goldman ([5, 7.5]). Let L be the (ungraded) cyclic Galois extension of K defined by the equation $X^p - X - T = 0$. Let $c \in k \setminus k^p$, where $p = \mathrm{char}\, k$. Let A be the crossed product defined by c, that is, A is generated over K by elements α,β satisfying the relations $\alpha^p - \alpha = X$, $\beta^p = c$, $\beta\alpha = (\alpha+1)\beta$. Let Λ be the crossed product constructed by the same relations over $k[T]$. Then Auslander and Goldman show that Λ is a nontrivial element of $\mathrm{Ker}(\mathrm{Br}(k[T] \to \mathrm{Br}(k))$. Hence A is not trivial in $\mathrm{Br}(k[T,T^{-1}])$, as $\mathrm{Br}(k[T]) \to \mathrm{Br}(k([T,T^{-1}])$ is a monomorphism. Also A cannot be split by a Galois extension of the form $l[T,T^{-1}]$. Indeed, otherwise $[\Lambda] \in \mathrm{Br}(l[T]/k[T])$, and then $[\Lambda] \in \mathrm{Im}(\mathrm{Br}(k) \to \mathrm{Br}(k[T]))$. Hence $[A] \notin \mathrm{Br}_g([T,T^{-1}])$, by the arguments given in IV.1.11. ■

IV.1.13 Example The most obvious example of the situation encountered in IV.1.11 is the case $K = \mathbf{R}[T,T^{-1}]$, where $\deg T = 1$. We leave it to the reader to show that

$$\mathrm{Br}_g(K_{(n)}) = \mathbf{Z}/2\mathbf{Z} = \{[\mathbf{H}[T,T^{-1}]],[\mathbf{R}[T,T^{-1}]]\}$$

if n is odd,

and

$$\mathrm{Br}_g(K_{(n)}) = \mathbf{Z}/2\mathbf{Z} \oplus \mathbf{Z}/2\mathbf{Z} = \{[\mathbf{R}[T,T^{-1}]], [\mathbf{H}[T,T^{-1}]], [\mathbf{C}[X,X^{-1},^{-}]], [\mathbf{C}[X,X^{-1},^{-},X^2=T]]\} \quad \text{if } n \text{ is even.}$$

IV.2 BRAUER GROUPS OF GR-LOCAL RINGS

Our first aim is to study splitting rings for gr-local rings. The class of gr-henselian rings, introduced in Chapter II, will play an important role. We start with a lemma.

IV.2.1 Lemma If R, m is a gr-henselian gr-local ring, and B is a possibly noncommutative graded finite R-algebra such that $R \subset Z(B)$, then the map $\mathrm{Idemp}(B_0) \to \mathrm{Idemp}((B/mB)_0)$ is surjective.

Proof: Take an idempotent $\bar{e} \in (B/mB)_0$, and lift $\bar{e}$ to $b \in B_0$. Then $A = R[b]$ is a finite commutative graded R-algebra, and $\bar{A}$ embeds in $\bar{B}$. By II.3.16, there exists $a \in \mathrm{Idemp}(A_0) \subset \mathrm{Idemp}(B_0)$ lifting $\bar{e}$. ■

IV.2.2 Proposition If R, m is a gr-local gr-henselian ring, and S is a finite commutative gr-local extension of R, then the map $\mathrm{Br}_g(S) \to \mathrm{Br}_g(S/mS)$ is a monomorphism. Consequently, the graded Brauer group of a strictly gr-henselian ring is trivial.

Proof: For any graded R-algebra A, let $\bar{A} = A/mA$. Suppose that A is a graded S-Azumaya algebra such that $\bar{A}$ is trivial in $\mathrm{Br}_g(\bar{S})$. By the results of Sec. 1, $\bar{A} = M_n(\bar{S})(\underline{d})$, for some integer n and $\underline{d} \in \mathbf{Z}^n$. Consider the matrix idempotent $f \in \bar{A}_0$, where all entries of f are zero, except for a one in the top left corner. By IV.1.2, f may be lifted to an idempotent $e \in A_0$. As S is gr-local, Ae is a graded free S-module. Representing A by left multiplication, we obtain a graded S-algebra morphism $\eta : A \to \mathrm{END}_S(Ae)$, which is clearly injective. Surjectivity of η follows from the fact that the induced map $\bar{\eta} : \bar{A}f = \bar{A}\bar{e} \to \overline{\mathrm{END}_S(Ae)} = \mathrm{END}_{\bar{S}}(\bar{A}f) \cong_g \bar{A}$ is surjective, using the graded version of Nakayama's lemma. So $A \cong_g \mathrm{END}_S(Ae)$, and $[A] = 1$ in $\mathrm{Br}_g(S)$. Finally, use IV.1.8. ■

IV.2.3 Lemma Let R be a gr-local ring, and A a finite graded R-algebra. Then A/R is separable if and only if $\bar{A}/\bar{R}$ is separable.

Proof: One implication is obvious. Conversely, if $\bar{A}/\bar{R}$ is separable, then we may prove that A is a graded direct summand of A^e. Let $m : A^e \to A$ be the canonical morphism, and $J = \mathrm{Ker} m$. Let $\delta : A \to J$ be defined by $\delta(a) = a \otimes 1 - 1 \otimes a$. By the separability of $\bar{A}$, $\bar{\delta}$ is inner, so there exists $\bar{x} \in (\bar{J})_0$ such that $\bar{\delta}\bar{a} = (\bar{\delta}\bar{a})\bar{x}$ for all $a \in A$. Choose a representative x of $\bar{x}$ in J_0. Since J is generated by δA, it follows from the graded version of Nakayama's lemma that $Jx = J$. The A^e-linear $\psi : J \to J : a \to ax$ is a graded isomorphism which can be extended to $\tilde{\psi} \in \mathrm{HOM}_{A^e}(A^e, J)$, by putting $\tilde{\psi}(a) = ax$. From $\tilde{\psi}^2 = \tilde{\psi}$, and after identification of $a - \tilde{\psi}(a)$ with $m(a - ax)$, it follows that $A^e \cong_g J \oplus A$ as graded A^e-modules. ■

IV.2.4 Theorem Let R be a gr-local ring, and A a graded R-Azumaya algebra.

(2.4.1) There exists a maximal commutative subalgebra S of A which is graded, separable, graded free as an R-module, and of the form $R[\beta]$, where $\beta \in A_0$. Furthermore, A is graded free as an S-module, and if $[A : R] = n^2$, then $\{1, \beta, \ldots, \beta^{n-1}\}$ is a basis for S as an R-module.

(2.4.2) There exists a Galois extension S_0 of R_0 such that $S = S_0 \otimes_{R_0} R$ splits A in $\mathrm{Br}_g(R)$.

(2.4.3) If R is quasi-strongly graded, then A may be split by a graded Galois extension S of R such that $\mathrm{Pic}_g(S) = 1$.

(2.4.4) If $R = R'_{(n)}$ for some quasi-strongly graded ring R', or if R is positively graded, or gr-henselian, then A may be split by a graded Galois extension S of R such that $\mathrm{Pic}^g(S) = 1$.

Proof: By IV.1.7, there exists a maximal commutative subalgebra W of $\bar{A}$, which is graded as an $\bar{R}$-algebra, separable, and of the form $W = \bar{R}[\alpha]$, $\alpha \in \bar{A}_0$. Lift α to an element $\beta \in A_0$, and let S be the graded free R-submodule of A generated by $\{1, \beta, \ldots, \beta^{n-1}\}$. Then $\{1, \beta, \ldots, \beta^{n-1}\}$ forms a basis of S, by an elementary application of the graded version of Nakayama's lemma. Let us show that $\beta^n \in S$; then it will follow that S is a graded subalgebra of A.

Consider a strict gr-henselization R^{sgh} of R (cf. II.3.23), and denote $\tilde{A} = A \otimes R^{\mathrm{sgh}}$. From 2.2, it follows that $\mathrm{Br}_g(R^{\mathrm{sgh}}) = 1$, so $\tilde{A} = M_n(R^{\mathrm{sgh}})(\underline{d})$, for some $\underline{d} \in \mathbf{Z}^n$. By the Cayley–Hamilton theorem, every element of $\tilde{A}$ satisfies its characteristic equation, which is a monic polynomial equation of degree n. In particular, $(\beta \otimes 1)^n = \beta^n \otimes 1 \in S \otimes R^{\mathrm{sgh}}$, so $\beta^n \in S$ since R^{sgh} is faithfully flat as an R-algebra. By IV.2.3 S is a separable R-algebra, so (2.4.1) is proved once we have verified that S is a maximal commutative subalgebra of A.

Observe that $A^S = S + (A^S \cap mA)$; indeed, $\bar{A}^{\bar{S}} = S$, so $A^S \subset \{x \in A : \bar{x}y = y\bar{x} \text{ for all } y \in \bar{S}\} = \{x \in A : \bar{x} \in \bar{S}\} = S + mA$. The canonical isomorphism $\eta : A^e \to \mathrm{END}_R(A)$ gives a commutative diagram of graded morphisms:

$$\begin{array}{ccc} A \otimes A^{\mathrm{opp}} & \xrightarrow{\eta} & \mathrm{END}_R(A) \\ \uparrow & & \uparrow \\ S \otimes A^{\mathrm{opp}} & \xrightarrow{\eta} & \mathrm{END}_S(A) \end{array}$$

Viewing the vertical monomorphisms as inclusions, we have that $A^S = A^S \otimes 1 \subset (\mathrm{END}_S(A))^{S \otimes A^{\mathrm{opp}}} \subset \mathrm{END}_R(A)^{S \otimes A^{\mathrm{opp}}} = (A \otimes A^{\mathrm{opp}})^{S \otimes A^{\mathrm{opp}}} = A^S \otimes 1$. By the double commutator theorem, we therefore have that $S \otimes A^{\text{opp}} = (\text{END}_S(A))^{A^S}$, so $(S \otimes A^{\text{opp}}) \otimes A^S = \text{END} = S(A)$, by [53, 2.13].

Finally, the monomorphism $S \to A^S$ is an isomorphism: A^S is a graded S-Azumaya algebra, and $A^S = S + (A^S \cap mA)$. As $A^S \cap mA$ is a graded two-sided ideal of A^S, there is a graded ideal I of S with $IA^S = A^S \cap mA$, and $IA^S \cap S = I$. Hence $I = A^S \cap mA \cap S = mA \cap S = mS$, and $A^S = S + (A^S \cap mA) = S + mSA^S = S + mA^S$; therefore $A^S = S$ by the graded version of Nakayama's lemma. So $S \otimes A^{\text{opp}} = \text{END}_S(A)$, and S splits A. This finishes the proof of (2.4.1).

As R is gr-local, R is connected. One of the direct summands S' of S already splits A; indeed, since R embeds in S, the identity element of R is an idempotent e in S. Consider $S' = eS$; then S' has no proper idempotents, and $(1 - e)S \cap R = 0$. Clearly S' splits A. Replace S by S'; then S is still of the form $R[\beta]$, where $\deg \beta = 0$. Using the embedding theorem ([23, II.2.9]), we may find an extension N_0 of $S_0 = R_0[\beta]$ such that N_0 is a Galois extension of R_0. Then $N = N_0 \otimes R$ is a graded Galois splitting ring for A. This proves (2.4.2).

If R is quasi-strongly graded, then R contains a homogeneous unit of positive degree. Indeed, $\bar{R} = k[T, T^{-1}]$, and T may be lifted to a homogeneous unit t. Suppose $\deg T = \deg t = d$; then $N[X]/(X^d - t)$ may be embedded in a graded Galois extension M of R, by II.2.15. Also M contains a unit of degree one, so M is strongly graded. Furthermore M is gr-semilocal, so $\text{Pic}_g(M) = 1$; this finishes the proof of (2.4.3). If $R = R'_{(n)}$ for some quasi-strongly graded ring R', then we may construct M as above. Now M is not necessarily strongly graded, but it is of the form $M'_{(n)}$ for some strongly graded ring M'. So this time $\text{Pic}^g(M) = 1$. In the gr-henselian case, the ring N constructed above is a gr-local ring, so $\text{Pic}^g(N) = 1$. Finally, if R is positively graded, then N is also positively graded, and $\text{Pic}^g(N) = 1$. ■

IV.2.5 Corollary If R is gr-local, then
(2.5.1) $\mathrm{Br}_g(R) = \varinjlim H^2(\mathrm{Gal}(S/R), U_0(S))$ if R is quasistrongly graded.
(2.5.2) $\mathrm{Br}_g(R) = \varinjlim H^2_{\mathrm{gr}}(\mathrm{Gal}(S/R), U(S))$ if $R = R'_{(n)}$ for some quasistrongly graded ring R', or if R is gr-henselian, or positively graded. The inductive limits are taken over all graded Galois extensions S of R.

Proof: Apply IV.2.4, III.5.9 and III.5.3.2. ■

IV.2.6 Remark IV.2.5 states that every graded Azumaya algebra over a gr-local ring satisfying one of the properties in IV.2.5 is equivalent to a crossed product. In the case of a quasistrongly graded ring, this crossed product may be taken over elements of degree zero (cf. Example IV.1.10). At present we do not know whether there exists a graded Azumaya algebra over a general gr-local ring which is not equivalent to a crossed product. Indeed, the Galois splitting ring constructed in IV.2.4 is only gr-semilocal, so it might have a nontrivial Pic^g. We have the following result.

IV.2.7 Proposition Every graded Azumaya algebra A over a gr-local ring is equivalent to a generalized crossed product in the sense of Kanzaki ([34]) or Chapter I.

Proof: Using IV.2.4 we obtain a graded Galois extension S of R splitting A. Consider the exact sequence III.5.5, where we replace Amitsur cohomology groups by Galois cohomology groups. $\beta_2([A])$ can be represented by a cocycle $I : G = \mathrm{Gal}(S/R) \to \mathrm{Pic}_g(S) : \sigma \to I_\sigma$. The action of G on $\mathrm{Pic}_g(S)$ is defined by letting $\sigma \cdot I$ be the R-module I with action from S given by $s \cdot x = \sigma^{-1}(s)x$. We then obtain graded S-isomorphisms $g_{\sigma,\tau} : I_\sigma \otimes I \to I_{\sigma\tau}$.

For $\sigma, \tau, \rho \in G$, consider the graded isomorphism $g_{\sigma\tau,\rho} \circ (g_{\sigma,\tau} \otimes 1) \circ (1 \otimes g^{-1}_{\tau,\rho}) \circ g^{-1}_{\sigma,\tau\rho}$ of $I_{\sigma\tau\rho}$, which is multiplication by a unit

$h(\sigma,\tau,\rho)$ of S. Then $h \in Z^3(G,U_0(S))$ represents $\gamma_2 \circ \beta_2([A])$, hence $[h] = 1$ in $H^3(G,U_0(S))$. As a consequence, we may choose the isomorphisms $g_{\sigma,\tau}$ in such a way that $h(\sigma,\tau,\rho) = 1$. Consider the graded R-Azumaya algebra B given by $B = \oplus_{\sigma\in G} I_\sigma$ and $(su)(tv) = s\sigma^{-1}(t)g_{\sigma,\tau}(u \otimes v)$ for $u \in I_\sigma$, $v \in I_\tau$, $s,t \in S$.

Then B is just the graded R-algebra used in the proof of III.5.5 to show exactness at $H^1(S/R,\mathrm{Pic}_g)$. So $\beta_2([B]) = [I]$; hence $A \otimes B^{\mathrm{opp}}$ is represented by a crossed product over a cocyle $f \in Z^2(G,U_0(S))$. Therefore A is equivalent to the generalized crossed product A' defined by $A' = \oplus_{\sigma\in G} I$ and $(su)(tv) = f(\sigma,\tau)s\sigma^{-1}(t)g_{\sigma,\tau}(u \otimes v)$ for $u \in I_\sigma$, $v \in I_\tau$, $s,t \in S$. ■

In IV.1.11, we have shown that for a perfect field k, ${}_n\mathrm{Br}(R[T,T^{-1}]) = {}_n\mathrm{Br}_g(R[T,T^{-1}, \deg T = n])$. Let us generalize this property to the case of a gr-local ring.

IV.2.8 Theorem Suppose that R is quasistrongly graded gr-local Krull domain such that n^{-1} is in R. Then ${}_n\mathrm{Br}(R) = {}_n\mathrm{Br}_g(R_{(n)})$.

Proof: The strict graded henselization of R is of the form $N[X,X^{-1}]$, where N is strictly henselian. In [24], Ford proved that such a ring has no n-torsion in its Brauer group. Therefore, given $[A] \in {}_n\mathrm{Br}(R)$, we can find a Galois extension S of the form $S = S_0[X,X^{-1}]$ of splitting A. Also, the Galois group G is of exponent n. Now, as $\mathrm{Pic}(S) = \mathrm{Pic}^g(S) = 1$ (S is a Krull domain), we have that ${}_n\mathrm{Br}(S/R) = {}_nH^2(G,U(S))$, and ${}_n\mathrm{Br}_g(S_{(n)}/R_{(n)}) = {}_nH^2_{\mathrm{gr}}(G,U(S_{(n)}) = {}_n\mathrm{Ker}(H^2(G,U(S_{(n)})) \to H^2(G,\mathbf{Z})) = {}_nH^2(G,U(S_{(n)}))$, because a unit in $U(S_{(n)})$ has degree in $n\mathbf{Z}$ and $H^2(G,n\mathbf{Z}) = 1$. Thus $\mathrm{Br}_g(S_{(n)}/R_{(n)}) = \mathrm{Br}(S/R)$, and the result follows if we take the union over all Galois extensions S of the form $S_0[X,X^{-1}]$. ■

As another application, let us show that the map $\mathrm{Br}_g(R) \to \mathrm{Br}_g(R/m)$ mentioned in IV.2.2 is also surjective, at least when R is gr-complete.

IV.2.9 Theorem If R is a gr-complete noetherian gr-local ring, then $\mathrm{Br}_g(R) \cong \mathrm{Br}_g(R/m)$.

Proof: Consider the canonical morphism $\mathrm{Br}_g(R) \to \mathrm{Br}_g(R/m)$, which is a monomorphism, by IV.2.2 and II.3.18. Suppose R is reduced. Let $\bar{A}$ be a graded $\bar{R}$-Azumaya algebra. Then $\bar{A}$ may be split by a Galois extension L of $\bar{R}$. The argument used in the proof of IV.2.4 shows that there exists a graded Galois extension S of R such that $\bar{S} = L$. S is gr-local, as it is a graded Galois extension of a gr-henselian ring. Since S is reduced and connected, all units of S are homogeneous. By (2.5.2) it suffices to show that $H^2_{\mathrm{gr}}(G, U(S)) \to H^2_{\mathrm{gr}}(G, U(L))$ is onto ($G = \mathrm{Gal}(L/\bar{R}) = \mathrm{Gal}(S/R)$). To this end, let $S_n = 1 + (m^n S)_0$; then we have an exact sequence

$$1 \longrightarrow S_1 \longrightarrow U(S) \longrightarrow U(L) \longrightarrow 1$$

resulting in a long exact sequence

$$\cdots \longrightarrow H^q(G, S_1) \longrightarrow H^q(G, U(S)) \longrightarrow H^q(G, U(L)) \longrightarrow \cdots$$

We claim that $H^q(G, S_1) = 1$, for all $q \geq 1$. As a G-module, the multiplicative group S_n/S_{n+1} is isomorphic to $(m^n S)_0/(m^{n+1}S)_0$. Since $m^n S/m^{n+1}S = m^n/m^{n+1} \otimes_{\bar{R}} \bar{S}$ and $H^q(G, m^n/m^{n+1}) = 0$, $H^q(G, m^n S/m^{n+1}S) = H^q(G, \bar{S}) = H^q(\bar{S}/\bar{R}, Id) = 0$, by [37, II.2.1]. Suppose $f \in Z^q(G, m^n S/m^{n+1}S)$, where $\deg f = 0$. Then there exists $g_i \in K^{q-1}(G, S_i)$ $(i = 1, 2, \ldots)$ such that $f\delta_{q-1}(g_1, \ldots, g_i) \in Z^q(G, S_{i+1})$. Then $g_1, g_1g_2, g_1g_2g_3, \ldots$ forms a $\mathcal{U}$-m-gr-Cauchy sequence; hence there exists $g \in K^{q-1}(G, S_1)$ such that $g = \lim g_1 \ldots g_m$, by gr-completeness. Furthermore, we have that $g(g_1, \ldots, g_m)^{-1} \in K^{q-1}(G, S_{m+1})$. Setting $h = f\delta_{q-1}g$, we have $h = f\delta(g_1,\ldots,g_m)\delta(g(g_1,\ldots,g_m)^{-1})$, while $f\delta(g_1,\ldots,g_m) \in Z^q(G,S_m)$ and $\delta(g(g_1,\ldots,g_m)^{-1}) \in B^q(G,S_{m+1})$. Consequently $h \in Z^q(G,S_m)$ for every $m \geq 1$, so $h = 1$. This shows that $H^q(G,S_1) = 1$, and $H^2(G,U(L)) = H^2(G,U(S))$. By III.5.11, $H^2_{\text{gr}}(G,U(S)) = \text{Ker}(H^2(G,U(S)) \to H^2(G,\mathbf{Z})) = \text{Ker}(H^2(G,U(L)) \to H^2(G,\mathbf{Z})) = H^2_{\text{gr}}(G,U(L))$, and this finishes the proof in the case where R is reduced. For the general case, we invoke IV.3.6. ■

We conclude this section with a generalization of IV.2.7, following an idea of DeMeyer [21].

IV.2.10 Theorem Let R be a gr-semilocal ring having no idempotents but 0 and 1. Then every graded R-Azumaya algebra may be split by a graded Galois extension of R. Consequently, every graded R-Azumaya algebra is equivalent to a generalized crossed product in the sense of Kanzaki.

Proof: Let $J = J^g(R) = m_1 \cap \cdots \cap m_n$, $\bar{R} = R/J = K_1 \oplus \cdots \oplus K_n$, where K_i is a graded field. Let $\bar{A} = A/JA$; then $\bar{A} = \bar{A}_1 \oplus \cdots \oplus \bar{A}_n$, where $\bar{A}_i$ is a gr-central simple K_i-algebra. As A is a graded R-algebra which is gr-projective as an R-module, A is graded free, and has a homogeneous basis consisting of elements of degree zero. Suppose that $[A : R] = m^2$; then $[\bar{A}_i : K_i] = m^2$, for every i. Using IV.1.7, we may find a maximal commutative subalgebra L_i of $\bar{A}_i$, which is graded and separable as a K_i-algebra, and of the form $L_i = K_i[\alpha_i]$, where $\deg \alpha_i = 0$. Also $[L_i : K_i] = m$.

Let $\alpha = \alpha_1 + \cdots + \alpha_n$. Then $\bar{R}[\alpha]$ is a maximal commutative subalgebra of $\bar{A}$, and $\bar{R}[\alpha]$ is separable over $\bar{R}$. Let P_i be the minimum polynomial of α_i over $(K_i)_0$; then α satisfies the monic polynomial $P = P_1 + \cdots + P_n \in \bar{R}_0[X]$. Lift P to a monic polynomial $Q \in R_0[X]$, and let $S = K[X]/(Q(X))$. Then S is a finitely generated graded free R-algebra, and S/R is separable. As $[A \otimes S]$ is in the kernel of the map $\mathrm{Br}^g(S) \to \mathrm{Br}^g(S/JS)$, we may assume that $[A]$ is in $\mathrm{Ker}(\mathrm{Br}_g(R) \to \mathrm{Br}_g(\bar{R}))$. Take $\beta \in A_0$ such that $\bar{\beta} = \alpha$, and let S be the graded free R-module with basis $1, \beta, \beta^2, \ldots, \beta^{m-1}$. As in the proof of IV.2.4, we may show that $\beta^m \in S$. But now it is more appropriate to use the graded completion rather than henselization. The rest of the proof goes exactly as in IV.2.4. ■

IV.3 THE BRAUER GROUP OF A GRADED RING MODULO A GRADED IDEAL

In the preceding section, we have shown that for a gr-complete gr-local ring, $\mathrm{Br}_g(R) \cong \mathrm{Br}_g(R/m)$. In this section, we present two similar results. We start with a very easy one.

IV.3.1 Proposition Let R be a positively graded ring; then $\mathrm{Br}_g(R) \cong \mathrm{Br}_g(R/R_+) \cong \mathrm{Br}_g(R_0)$, where R_0 is considered to be a trivially graded ring.

Proof: From III.3.13, it follows that we have an equivalence between the categories $\underline{\underline{FP}}_g(R)$ and $\underline{\underline{FP}}_g(R/R_+)$ given by the functors $\cdot \otimes_{R_0} R$ and $\cdot \otimes_R (R/R_+)$. As a graded Azumaya algebra is also a faithfully projective module, and as these functors send graded Azumaya algebras to graded Azumaya algebras, this also establishes an equivalence between $\underline{\underline{Az}}_g(R)$ and $\underline{\underline{Az}}_g(R/R_+)$. Finally, use the definition III.4.1. ■

IV.3.2 Note In Chapter V, we shall show that Br and Br_g coincide for a trivially graded ring. It will therefore follow that $\mathrm{Br}_g(R) \cong \mathrm{Br}(R_0)$ for a positively graded ring.

IV.3.3 Proposition Let I be a nilpotent graded ideal of R, and denote $\bar{R} = R/I$. Then for every finitely generated graded $\bar{R}$-module $\bar{P}$, there exists a graded projective R-module P such that $P \otimes_R \bar{R} \cong_g \bar{P}$. If P is a finitely generated graded projective R-module, and if $P \otimes_R \bar{R}$ is graded free, then P is graded free. Consequently $\mathrm{Pic}_g(R) = \mathrm{Pic}_g(\bar{R})$ and $K_0^g(R) = K_0^g(\bar{R})$.

Proof: As $\bar{P}$ is a direct summand in $\bar{R}^n(\underline{d})$, there exists a (graded) projection $\bar{R}^n(\underline{d}) \to \bar{P}$, represented by a homogeneous idempotent of $M_n(\bar{R})(\underline{d})$. This may be lifted to a homogeneous idempotent of $M_n(R)(\underline{d})$, giving a projection $R^n(\underline{d}) \to P$, for some graded projective R-module P. Clearly $P/IP \cong_g \bar{P}$. This yields the first assertion. Suppose $\bar{P} \cong_g \bar{R}^n(\underline{d})$, and consider the matrix idempotents in $M_n(\bar{R})(\underline{d})$ given by the projections $\bar{R}^n(\underline{d}) \to \bar{R}(d_i)$. Lift these to homogeneous projections p_i in $\mathrm{END}_R(P)$. Then $P \cong_g p_1(P) \oplus \cdots \oplus p_n(P)$, and it follows easily that $p_i(P) \cong_g R(d_i)$. The final statement now follows immediately. ∎

IV.3.4 Proposition Let I be a nilpotent graded ideal of R; then the map $\mathrm{Br}_g(R) \to \mathrm{Br}_g(R/I)$ is surjective. If $\delta(R) < +\infty$ (cf. III.2.5), then $\mathrm{Br}_g(R) \cong \mathrm{Br}_g(R/I)$.

Proof: As Verschoren remarks in his paper [71], this proof is an easy adaptation of a similar ungraded proof, given by DeMeyer in [22]; see also [33]. The condition $\delta(R) < +\infty$ appears at the point where one has to invoke III.3.18. ∎

IV.3.5 Theorem If $\delta(R) < +\infty$, then $\mathrm{Br}_g(R) = \mathrm{Br}_g(R_{\mathrm{red}})$.

Proof: If R is noetherian, the statement follows from IV.3.4, as the nilradical of R is nilpotent in this case. So let R be non-noetherian and denote $R_{\mathrm{red}} = S$, $\pi : R \to S$, $f : \mathrm{Br}_g(R) \to \mathrm{Br}_g(S)$. Take a graded R-Azumaya algebra A, and suppose that $f([A]) = 1$. Then there exists a noetherian graded subring R' of

R, and a graded R'-Azumaya algebra such that $A' \otimes_{R'} R \cong_g R$. Let $S' = R'_{\text{red}}$, and suppose $f'([A']) \neq 1$. Then there exists a noetherian graded subring S'' of S which splits A'. Let $R'' = \{x \in R : \pi(x) \in S''\}$, $A'' = A' \otimes_{R'} R''$; then $f''([A'']) = 1$. Consequently $[A''] = 1$, and $[A] = 1$. Hence f is injective. To show that f is surjective, take a graded S-Azumaya algebra B. Then there exists a noetherian graded subring S' of S and a graded S'-Azumaya algebra B' such that $B' \otimes_{S'} S = B$. Let $R' = \{x \in R : \pi(x) \in S'\}$; then $R'_{\text{red}} = S'$ so there exists a graded R'-Azumaya algebra A' such that $f'([A']) = B'$. Let $A = A' \otimes_{R'} R$; then $f([A]) = [B]$. ∎

IV.4 BRAUER GROUPS OF REGULAR GRADED RINGS

Let R be a commutative graded ring. We say that R has *gr-global dimension* $\leq n$ if every graded R-module has a finite projective resolution in R-gr of length smaller than or equal to n. It may be established easily that this is equivalent to the fact that for every $A, B \in R$-gr, $\text{EXT}^{n+1}(A, B) = 0$. R is said to be *gr-regular* if the gr-global dimension gr gl dim R is finite, and if R is noetherian.

IV.4.1 Lemma Let R be a commutative graded ring; then R is regular if and only if R is gr-regular. Moreover, gr gl dim $R \leq$ gr gl dim $R + 1$.

Proof: Cf. [48, 49]. ∎

Also recall from [48, 49] that for a regular graded domain, the gr-Krull dimension and the gr-global dimension coincide. It is straightforward to show the following:

IV.4.2 Theorem Let R be a regular graded domain, and let $K^g = Q^g(R)$. Then the canonical map $\mathrm{Br}_g(R) \to \mathrm{Br}_g(K^g)$ is a monomorphism.

Proof: Consider the following diagram of abelian groups:

$$\begin{array}{ccccc} \mathrm{Br}_g(R) & \xrightarrow{\alpha} & \mathrm{Br}_g(K^g) & & \\ \downarrow{\scriptstyle\beta} & & \downarrow{\scriptstyle\gamma} & & \\ \mathrm{Br}(R) & \xrightarrow{\delta} & \mathrm{Br}(K^g) & \xrightarrow{\varepsilon} & \mathrm{Br}(K) \end{array}$$

In V.3, we will show that β and γ are monomorphisms. From [53, 6.19], it follows that $\varepsilon \circ \delta$ is a monomorphism, so δ is a monomorphism. Hence α is a monomorphism. ∎

Our next aim is to show that for a regular graded domain of gr-global dimension at most two, $\mathrm{Br}_g(R)$ may be written as the intersection of the graded Brauer groups of the graded localizations at all minimal nonzero graded prime ideals. We need some preliminary results first.

IV.4.3 Lemma Let M be a graded left R-module, where $\mathrm{gr\ gl\ dim}\, R \leq n$, and if there exists an exact sequence in R-gr of the form

$$0 \longrightarrow X \longrightarrow P_{n-1} \longrightarrow P_{n-2} \longrightarrow \cdots \longrightarrow P_0 \longrightarrow M \longrightarrow 0$$

where each P_i is projective, then X is projective.

Proof: Exactly as in [51, p. 135].

IV.4.4 Lemma Let R be a graded regular domain, and suppose $\mathrm{gr\ gl\ dim}\, R \leq 2$. Then every finitely generated graded reflexive R-module M is projective.

Proof: Take an exact sequence in R-gr

$$0 \longrightarrow L \longrightarrow F \longrightarrow M^* \longrightarrow 0$$

where F is graded free and of finite rank. Taking duals, we obtain an exact sequence

$$0 \longrightarrow M^{**} = M \longrightarrow F^* \longrightarrow Q \longrightarrow 0$$

where Q is a graded submodule of L^*. Hence Q is torsion free, so $Q \subset K^g \otimes Q$. Take a finite homogeneous basis of $K^g \otimes Q$ over K^g. Expressing the homogeneous generators of Q in terms of this basis, we can construct an embedding $Q \subset G$, where G is a graded free R-module of finite rank. So we have an exact sequence

$$0 \longrightarrow Q \longrightarrow G \longrightarrow G/Q \longrightarrow 0$$

and therefore

$$0 \longrightarrow M \longrightarrow F^* \longrightarrow G \longrightarrow G/Q \longrightarrow 0$$

is also exact. From the preceding lemma, it follows that M is projective. ∎

IV.4.5 Lemma (Graded Version of Krull's Principal Ideal Theorem). Let x be a homogeneous noninvertible element of a noetherian graded ring R. Then x is contained in a graded prime ideal of height one in $\mathrm{Spec}^g(R)$.

Proof: From the classical version of Krull's principal ideal theorem (cf., for example, [35, Theorem 142]), it follows that $x \in p$, where p is a prime ideal of height one, that is, we have a minimal chain of prime ideals $p_0 \subsetneqq p$ in $\mathrm{Spec}(R)$. Then p_0 is a graded prime ideal, because it is a minimal prime ideal.

Suppose that $x \notin p_0$, and consider p_g, the graded prime ideal generated by the homogeneous elements of p. Then $p_0 \neq p_g$, because $x \in p_g$, so $p_g = p$, and the result follows from the fact that $p = p_g$ is graded. If $x \in p_0$, then take an arbitrary homogeneous element $y \in R \backslash p_0$ which is not invertible, and apply the preceding result to y. ∎

IV.4.6 Lemma Let R be a commutative noetherian graded ring, and let A be a central graded R-algebra which is faithfully projective as an R-module. If $Q_p^g(A)$ is a (graded) Azumaya algebra for every graded prime of height one in R-gr, then A is a graded Azumaya algebra.

Proof: It suffices to show that $\eta : A^e \to \mathrm{END}_R(A)$ is a graded isomorphism, or, after graded localization, that for every $m \in \mathrm{Max}^g(R)$, $Q_m^g(\eta) : Q_m^g(A) \otimes Q_m^g(A)^{\mathrm{opp}} \to \mathrm{END}(Q_m^g(A))$ is an isomorphism. By a rank argument, this is the case if $\det Q_m^g(\eta)$ lies in $U_0(R)$. But if not, then $\det Q_m^g(\eta)$ lies in a prime ideal of height one in R-gr, by IV.4.5, contradicting the hypothesis. ■

IV.4.7 Let us recall some basic facts about graded orders, referring to the monograph [42] for an extensive discussion. Let R be a regular graded domain, or, more generally, a Krull domain. Denote $Q^g(R) = K^g = k[T,T^{-1}]$, and $Q(R) = K = k(T)$. Let $\sum^g$ be a gr-central simple K^g-algebra, and let A be a graded R-algebra contained in $\sum^g$, such that A contains a homogeneous K^g-basis of $\sum^g$, and such that every element of A is integral over R. Then we say that A is a *gr-order* in $\sum^g$. It follows immediately that in this case A is an order in $K \otimes \sum^g = \sum$. Conversely, let $\sum$ be a central simple K-algebra, and suppose that A is an R-order in $\sum$ which is graded as an R-algebra. It may be shown that $K^g \otimes A = \sum^g$ is a gr-central simple K^g-algebra, and that A is a gr-order in $\sum^g$. If A is not contained properly in any other gr-order, then we call A a *gr-maximal order*. It may be shown that in this case, A is a maximal order in $\sum$. Then it follows from [53, 6.31] that $M_n(A)(\underline{d})$ is a maximal order in $M_n(\sum)$; hence it is a gr-maximal order in $M_n(\sum^g)(\underline{d})$.

IV.4.8 Lemma Let R be a discrete gr-valuation ring, A a graded R-Azumaya algebra, and B a gr-maximal order in a gr-central simple K^g-algebra $\sum^g$. Then $A \cong_g B$.

Proof: Recall from II.1.15.3 that every graded ideal of a discrete gr-valuation ring is principal. Consider $c = \{x \in \sum^g : xB \in A\}$, which is a left graded A-ideal. From lemma IV.4.9 below, it follows that $c = At$ for some $t \in h(A)$. Now t is a unit in $\sum^g$, since $c \cap R \neq \Phi$ (we leave it to the reader to show this). From $cB \subset c$, it follows that $tB \subset At$, and $B \subset t^{-1}At$, so $B = t^{-1}At$ by the maximality of B. ■

IV.4.9 Lemma Let A be a graded R-Azumaya algebra, where R is a discrete gr-valuation ring. Then every left graded A-ideal is principal.

Proof: As the proof of 6.32 in [53]. ■

IV.4.10 Theorem If R is a graded regular domain of gr-global dimension at most two, then $\mathrm{Br}_g(R) = \cap \mathrm{Br}_g(Q_p^g(R))$, where the intersection runs over all graded primes of height one in $\mathrm{Spec}^g(R)$.

Proof: From IV.4.2 we know that there exist inclusions $\mathrm{Br}_g(R) \subset \mathrm{Br}_g(Q_p^g(R) \subset \mathrm{Br}_g(K^g)$, and this allows us to take intersection over all p of height one. Of course, $\mathrm{Br}_g(R) \subset \cap \mathrm{Br}_g(Q_p^g(R))$. Let Δ be a graded skewfield with center K^g, lying in the image of $\mathrm{Br}_g(Q_p^g(R))$, for each graded prime of height one. Then for each p of height one in R-gr, there exists a graded R_p^g-Azumaya algebra $A(p)$ such that $K^g \otimes A(p) \cong_g M_{n(p)}(\Delta)(\underline{d})$, for some $n(p) \in \mathbf{N}_0$, $\underline{d} \in \mathbf{Z}^{n(p)}$. Let A be a gr-maximal order in Δ. (For the existence of gr-maximal orders, see [42].) Then $\Delta = K \otimes A$. We are done if we can prove that A is Azumaya. As a maximal order, A is reflexive, so A is projective, by IV.4.4. Furthermore $Z(A) = R$, as $Z(A) \subset Z(\Delta) = K^g$, and every element of $Z(A)$ is integral over R. Finally, A_p^g is a gr-maximal R_p^g-order in Δ. So $M_{n(p)}(A_p^g)(\underline{d})$ is a gr-maximal R_p^g-order in $M_{n(p)}(\Delta)(\underline{d})$ by IV.4.7. From IV.4.8, it follows that $M_{n(p)}(A_p^g)(\underline{d}) \cong_g A(p)$, so $M_{n(p)}(A_p^g)(\underline{d})$ and A_p^g are R_p^g-Azumaya algebras. From IV.4.6, it follows now that A is a graded R-Azumaya algebra. ■

IV.4.11 Remark In particular, when R is a gr-Dedekind domain, we have the following:

$$\mathrm{Br}_g(R) = \cap \mathrm{Br}_g(R_p^g) \subset \mathrm{Br}(R) = \cap \mathrm{Br}(R_p)$$

where the intersections are, respectively, taken over all elements of height one in $\mathrm{Spec}^g(R)$ and $\mathrm{Spec}(R)$.

Chapter V

Etale Cohomology for Graded Rings

V.1 COHOMOLOGY ON THE GR-ETALE SITE

V.1.1 Let us recall some generalities about homological algebra; for details, we refer to the literature, for example, [8, 17]. Let $\underline{\mathcal{C}}$ be an abelian category having enough injectives, and $f : \underline{\mathcal{C}} \to \underline{\mathcal{D}}$ a left exact functor, where $\underline{\mathcal{D}}$ is another abelian category. Then it is a well-known fact that we have an essentially unique sequence of functors $R^i f : \underline{\mathcal{C}} \to \underline{\mathcal{D}}$, for $i \geq 0$, called the *right derived functors* of f. These functors satisfy the following properties:

(1.1.1) $R^0 f = f$.

(1.1.2) $R^i f(I) = 0$ if $i \geq 0$ and if I is an injective object of $\underline{\mathcal{C}}$.

(1.1.3) For any exact sequence $0 \to A \to B \to C \to 0$ in $\underline{\mathcal{C}}$, we have a long exact sequence in $\underline{\mathcal{D}}$ as follows: $0 \to f(A) \to$

$f(B) \to f(C) \to R^1f(A) \to R^1f(B) \to R^1f(C) \to R^2f(A) \to R^2f(B) \to \cdots$.
Moreover, the association of this long exact sequence to the short exact sequence behaves functorially.

The functors R^if may be determined as follows: take an object $A \in \underline{C}$, and let $P \to A \to X^0 \to X^1 \to X^2 \to \cdots$ be an injective resolution of A. Then we obtain the following complex in $\underline{D} : 0 \to f(X^0) \to f(X^1) \to f(X^2) \to \cdots$. Then $R^if(A)$ is defined as the ith cohomology object of this sequence; one may easily check that $R^if(A)$ is independent of the chosen resolution, and that R^if satisfies (1.1.1–3). Also, the uniqueness is easily verified: let R^if and S^if be two sequences of functors satisfying (1.1.1–3). Then take an injective object B of $\underline{C}$, and write the long exact sequence associated to $0 \to A \to B \to C \to 0$. It is easily seen that $R^if(A) \cong S^if(A)$.

Recall that an object A of $\underline{C}$ is called f-*acyclic* if $R^if(A) = 0$ for all $i > 0$. Let us present a generalization of this. Suppose $\underline{D} = \underline{Ab}$, the category of abelian groups.

V.1.2 Proposition Suppose that in V.1.1., $\underline{D} = \underline{Ab}$, the category of abelian groups. Then to every morphism $d : B \to C$ in $\underline{C}$, we may associate a sequence of abelian groups denoted $R^if(d) = R^i_Cf(B)$ satisfying the following properties:
(1.2.1) $R^0f(d) = f(\ker d)$.
(1.2.2) We have a long exact sequence $0 \to f(\ker d) \to f(B) \to f(C) \to R^1f(d) \to R^1f(B) \to R^1f(C) \to R^2f(d) \cdots$.
The abelian groups $R^if(d)$ are unique up to isomorphism.

Note V.1.2 is a generalization of V.1.1 in the following sense: V.1.1 associates a long exact sequence to $0 \to A \to B \to C \to 0$, while V.1.2 associates one to $0 \to A \to B \to C$.

Proof: Consider injective resolutions Y and Z of B and C. We then obtain a map between complexes

$$\begin{array}{ccccccccc} 0 & \longrightarrow & f(Y_0) & \xrightarrow{\Delta_0} & f(Y_1) & \xrightarrow{\Delta_1} & f(Y_2) & \xrightarrow{\Delta_2} & \cdots \\ & & \downarrow d_0 & & \downarrow d_1 & & \downarrow d_2 & & \\ 0 & \longrightarrow & f(Z_0) & \xrightarrow{D_0} & f(Z_1) & \xrightarrow{D_1} & f(Z_2) & \xrightarrow{D_2} & \cdots \end{array}$$

As in III.5, we obtain a new complex (W, ∇) as follows:

$$W_n = f(Y_n) \times f(Z_{n-1})$$

$$\nabla_n(y_n, z_{n-1}) = (\Delta_n y_n, D_{n-1} z_{n-1} - d_n y_n)$$

Let $R^i f(d) = \ker \nabla_i / \operatorname{im} \nabla_{i-1}$. The exact sequence (1.2.2) is just III.5.2. The uniqueness of $R^i f(d)$ follows from the uniqueness of $R^i f(B)$ and $R^i f(C)$, and the five-lemma. ■

V.1.3 Corollary With notations as above, we have for all $n \geq 1$.

(1.3.1) If C is f-acyclic, then $R^n_C f(B) = R^n f(B)$.

(1.3.2) If B is f-acyclic, then $R^n_C f(B) = R^{n-1} f(C)$.

(1.3.3) If d is an epimorphism in $\underline{\mathcal{C}}$, then $R^n_C f(B) = R^n f(A)$, where $A = \ker d$.

Proof: Use (1.2.2). ■

We conclude this homological intermezzo with the following proposition, taken from [72].

V.1.4 Proposition Let $\underline{\mathcal{B}}$, $\underline{\mathcal{C}}$, $\underline{\mathcal{D}}$ be abelian categories, and suppose that $\underline{\mathcal{B}}$ and $\underline{\mathcal{C}}$ have enough injectives; let $f : \underline{\mathcal{B}} \to \underline{\mathcal{C}}$ and $g : \underline{\mathcal{C}} \to \underline{\mathcal{D}}$ be left-exact functors such that f takes injective objects of $\underline{\mathcal{B}}$ to g-acyclic objects. Consider $A \in \underline{\mathcal{B}}$, and an injective resolution

$$(1.4.1) \quad 0 \longrightarrow A \longrightarrow X^0 \longrightarrow X^1 \longrightarrow \cdots$$

Let C_q be the object of $\underline{C}$ defined by

$$(1.4.2) \quad C_q = \mathrm{Ker}(f(X^q) \longrightarrow f(X^{q+1})) = f(\mathrm{Ker}(X^q \longrightarrow X^{q+1}),$$

for $q \geq 0$. Then we have long exact sequences

$$\begin{aligned}(1.4.3) \quad 0 &\longrightarrow R^1g(C_q) \longrightarrow R^{q+1}gf(A) \longrightarrow g(R^{q+1}f(A)) \\ &\longrightarrow R^2g(C_q) \longrightarrow R^1g(C_{q+1}) \longrightarrow R^1g(R^{q+1}f(A)) \\ &\longrightarrow \cdots \\ &\longrightarrow R^{p+1}g(C_q) \longrightarrow R^pg(C_{q+1}) \longrightarrow R^pg(R^{q+1}f(A))\end{aligned}$$

for every $q \geq 0$. Combining the sequences $q = 0$ and $q = 1$, we obtain the so-called *Leray spectral sequence*

$$\begin{aligned}(1.4.4) \quad 0 &\longrightarrow R^1g(f(A)) \longrightarrow R^1gf(A) \longrightarrow g(R^1f)(A) \\ &\longrightarrow R^2g(f(A)) \longrightarrow \mathrm{Ker}(R^2(gf)(A) \longrightarrow g(R^2f(A))) \\ &\longrightarrow (R^1g)(R^1f)(A) \longrightarrow R^3g(f(A))\end{aligned}$$

Proof: (Cf. Theorem 6.1 in [72].) Let $B_q = \mathrm{Im}(f(X^{q-1}) \rightarrow f(X^q))$. Then we have exact sequences

$$(1.4.5) \quad 0 \longrightarrow C_q \longrightarrow f(X^q) \longrightarrow B_{q+1} \longrightarrow 0$$

in $\underline{C}$. Applying g, we obtain a long exact sequence in $\underline{D}$. As $f(X^q)$ is acyclic, $R^ng(f(X^q)) = 0$ for $n > 0$, so we have

$$(1.4.6) \quad 0 \longrightarrow g(C_q) \longrightarrow gf(X^q) \longrightarrow g(B_{q+1}) \longrightarrow R^1g(C_q) \longrightarrow 0$$

$$(1.4.7) \quad R^pg(B) \cong R^{p+1}g(C_q)$$

for $p \geq 1$. Now we have another exact sequence in $\underline{C}$:

$$(1.4.8) \quad 0 \longrightarrow B_{q+1} \longrightarrow C_{q+1} \longrightarrow R^{q+1}f(A) \longrightarrow 0$$

Applying g, we obtain the following exact sequence in $\underline{D}$.

$$\begin{aligned}0 &\longrightarrow g(B_{q+1}) \longrightarrow g(C_{q+1}) \longrightarrow g(R^{q+1}f(A)) \\ &\longrightarrow R^1g(B_{q+1}) \cong R^2g(C_q) \longrightarrow R^1g(C_{q+1}) \longrightarrow \cdots\end{aligned}$$

We have a commutative diagram

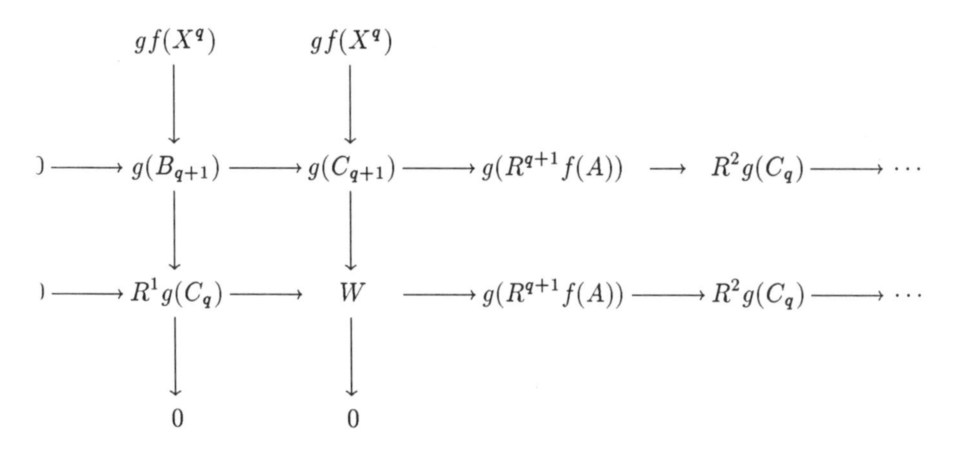

in which the first column is (1.4.6), while the second is just the definition of W as the cokernel of the composite $gf(X^q) \to g(B_{q+1}) \to g(C_{q+1})$. A little diagram chasing shows that the exactness of the first of the long exact sequences implies the exactness of the second. Finally, we can identify W with $R^{q+1}(gf)(A)$: $C_{q+1} = \mathrm{Ker}(f(X^{q+1}) \to f(X^{q+2}))$ and, by the left exactness of g, $g(C_{q+1}) = \mathrm{Ker}(gf(X^{q+1}) \to gf(X^{q+2}))$, so $\mathrm{Coker}(gf(X^q) \to g(C_{q+1})) = \mathrm{Ker}(gf(X^{q+1}) \to gf(X^{q+2}))/\mathrm{Im}(gf(X^q) \to gf(X^{q+1})) = R^{q+1}(gf)(A)$. The second row of the diagram is now just (1.4.3). We leave it to the reader to derive the exact sequence (1.4.4). ■

V.1.5 We return to graded rings. Let E_g be a class of morphisps of graded rings, satisfying the following conditions:

(1.5.1) All graded isomorphisms are in E_g.

(1.5.2) The composite of two morphisms of E_g is in E_g.

(1.5.3) If $f : R \to S$ is in E_g, and R' is a commutative graded R-algebra, then $R' \to S \otimes R'$ is in E_g.

A morphism belonging to E_g will be called an E_g-morphism. W may take, for example, $E_g = (\text{ét}_g)$, (fl_g), (Zar_g), respectively, the classes of graded étale morphisms, graded flat morphisms, and graded open immersions.

Now fix a (graded) base ring R, and let E_g/R be the category of commutative graded R-algebras S for which the structural morphism $R \to S$ belongs to E_g, taking the E_g-morphisms for the morphisms. In the examples cited above, these categories will be denoted $(R_{\text{ét}_g})$, (R_{f1_g}), (R_{Zar_g}).

These definitions are taken from [44, ch. II], where we restricted to affine schemes, translated into ring theoretical language, and then stated the graded analog of the definition. In the sequel, we shall also use the ungraded counterpart. Thus we obtain the categories $(R_{\text{ét}})$, (R_{fl}), (R_{Zar}), which are called the *(small) étale, flat, and Zariski sites* on R. Note that $(R_{\text{ét}_g})$, (R_{fl_g}), (R_{Zar_g}) are still sites in the sense of [4, exp. 2]. We shall call these categories the *graded étale, flat and Zariski sites* on R.

Forgetting gradations, we obtain some morphisms of sites: $(R_{\text{ét}_g}) \to (R_{\text{ét}})$, $(R_{\text{fl}_g}) \to (R_{\text{fl}})$, and $(R_{\text{Zar}_g}) \to (R_{\text{Zar}})$. If an E_g-morphism is faithfully flat, then we call it an E_g*-covering.*

V.1.6 A covariant functor from E_g/R to the category of abelian groups will be called a *presheaf (of abelian groups)* on E_g/R. Indeed, a covariant functor on E_g/R corresponds to a contravariant functor—or presheaf—on the Grothendieck topology on $\text{Spec}^g(R)$ formed by the E_g-coverings; we refer to [4] and [44] for details. In the sequel, we shall state all our results in ring theoretical language.

If a presheaf P on E_g/R satisfies the conditions (1.6.1) and (1.6.2) below, then we call P a *sheaf.* Of course, this corresponds to the classical notion of a sheaf.

(1.6.1) If $s \in P(S)$, and $f : S \to S'$ is an E_g-covering, and if $P(f)(s) = 0$, then $s = 0$.

(1.6.2) If $S \to S'$ is an E_g-covering, and $s' \in P(S')$ and if $P(\varepsilon_1)(s') = P(\varepsilon_2)(s')$, where $\varepsilon_i : S' \to S' \otimes_S S'$ are given by

$\varepsilon_1(s) = 1 \otimes s$, $\varepsilon_2(s) = s \otimes 1$, then there exists $s \in P(S)$ such that $s' = P(f)(s)$.

For $E_g = (\text{ét}_g), (\text{fl}_g), (\text{Zar}_g)$, the categories of presheaves and sheaves on E_g/R are abelian categories having enough injectives. The proof is omitted here; actually, it is identical to the corresponding proof for E in [44]. These categories will be denoted $\underline{P}(E_g/R)$ and $\underline{S}(E_g/R)$.

A sequence of presheaves $0 \to F \to G \to H \to 0$ is exact in $\underline{P}(E_g/R)$ if and only if for every S in E_g/R, $0 \to F(S) \to G(S) \to H(S) \to 0$ is exact as a sequence of abelian groups. In the category $\underline{S}(E_g/R)$, this property holds only for exact sequences of the form $0 \to F \to G$. A map $r : G \to H$ is an epimorphism in $\underline{S}(R_{\text{ét}_g})$ if and only if for every S in $R_{\text{ét}_g}$, and for every $h \in H(S)$, there exists an E_g-covering $S \to S'$ such that the image of h in $H(S')$ lies in $\operatorname{Im} f(S')$, where of course $f(S')$ is the map $G(S') \to H(S')$. Concerning the graded étale site, a sequence $0 \to F \to G \to H \to 0$ is exact in $\underline{S}(R_{\text{ét}_g})$ if and only if for every $p \in \operatorname{Spec}^g(R)$, the sequence $0 \to F(R_p^{\text{sgh}}) \to G(R_p^{\text{sgh}}) \to H(R_p^{\text{sgh}}) \to 0$ is exact. The functors F, G, H may be defined on the strict graded henselization R_p^{sgh}, as this is a limit of gr-étale extensions of R (cf. Ch. II). The statements above are straightforward generalizations of [44, II.2.15]. The inclusion functor $\underline{S}(R_{\text{ét}_g}) \to \underline{P}(R_{\text{ét}_g})$ has a left adjoint functor $a : \underline{P}(R_{\text{ét}_g}) \to \underline{S}(R_{\text{ét}_g})$, which is exact and preserves inverse limits.

V.1.7 Cohomology Let $F, G \in \underline{S}(E_g/R)$, and d a morphism $F \to G$. Consider the functor $\Gamma : \underline{S}(E_g/R) \to \underline{Ab}$, given by $F \to F(R)$ (i.e., we take global sections). We define the E_g-cohomology groups of F by

$$H^n(E_g/R, F) = R^n\Gamma(F)$$

$$H^n_G(E_g/R, F) = R^n_G\Gamma(F)$$

Using the results of III.1, we may obtain that U_0 and U are sheaves on $R_{\text{ét}_g}$, R_{fl_g}, R_{Zar_g}. By III.3.9, the same conclusion holds for gr. In fact, for a reduced ring, gr is the sheaf associated to the constant presheaf $\mathbf{Z}$.

V.1.8 Definition A commutative graded ring is called *strongly graded on the E_g-site* if it satisfies one of the following equivalent conditions:

(1.8.1) R has an E_g-covering which contains a unit of degree one.
(1.8.2) The sequence $0 \to U_0 \to U \to \text{gr} \to 0$ is exact in $\underline{S}(E_g/R)$.
(1.8.3) The sheaf $a\,\text{Pic}_g$ associated to the presheaf Pic_g on E_g/R is zero.

Proof: Consider the exact sequence $0 \to U_0 \to U \to \text{gr} \to \text{Pic}_g \to \text{Pic}^g \to 0$ in $\underline{P}(E_g/R)$. Taking associated sheaves, we obtain an exact sequence in $\underline{S}(E_g/R)$, namely, $0 \to U_0 \to U \to \text{gr} \to a\,\text{Pic}^g = 0$, since Pic^g disappears for gr-local rings. $\text{Pic}_g(R)$, however, is trivial if and only if $\text{Pic}^g(R)$ is trivial and R contains a unit of degree one, as is easily checked. The result therefore follows. ∎

Observe that a strongly graded ring on the graded Zariski site is just a strongly graded ring in the classical sense. A strongly graded ring on the graded étale site is a quasistrongly graded ring in the sense of Chapter II. A strongly graded ring on the graded flat site is a ring for which all graded residue class fields are nontrivially graded. Observe that there exist graded rings which are not strongly graded on any of the three sites cited above, for example, positively graded rings, or rings of type $R[X,Y]$, where $\deg X > 0$ and $\deg Y < 0$.

V.1.9 Corollary If R is strongly graded on the E_g-site, then $H^n_{\text{gr}}(E_g/R, U) \cong H^n(E_g/R, U_0)$.

One of the main objects of this chapter is to construct a monomorphism $\mathrm{Br}_g(R) \to H^2_{\mathrm{gr}}(R_{\text{ét}_g}, U)$. In order to do so, we need an explicit description of the cohomology groups on the graded étale site. To this end, we introduce the following notation, for a morphism $d : F \to G$ in $\underline{S}(R_{\text{ét}_g})$:

$$\check{H}^n(R_{\text{ét}_g}, F) = \varinjlim H^n(S/R, F)$$

$$\check{H}^n_g(R_{\text{ét}_g}, F) = \varinjlim H^n_G(S/R, F)$$

where the limits are taken over all graded étale coverings of R, and the cohomology groups on the right-hand side are Amitsur cohomology groups, as considered in Chapter III. The inductive limits are well defined, as follows from an argument in [37]. In the case of a quasistrongly graded rings, Amitsur cohomology happens to be sufficient to describe gr-étale cohomology:

V.1.10 Proposition Let R be a graded ring which is strongly graded on the graded étale site. Then there exist canonical isomorphisms $H^n(R_{\text{ét}_g}, F) \cong \check{H}^n(R_{\text{ét}_g}, F)$ and $H^n_G(R_{\text{ét}_g}, F) \cong \check{H}^n_G(R_{\text{ét}_g}, F)$.

Proof: For any injective sheaf P, P is also injective as a presheaf. But then for every (graded) étale covering S of R, $H^n(S/R, P) = 0$, for $n > 0$ (cf. [44, III.2.4]). It follows that $\check{H}^n(R_{\text{ét}_g}, P) = 0$. Next consider an exact sequence $0 \to H \to F \to G$ in $\underline{S}(R_{\text{ét}_g})$. Using the graded version of Artin's refinement theorem (II.4.17), we may associate the following long exact sequence to this:

$$0 \longrightarrow H(R) \longrightarrow F(R) \longrightarrow G(R) \longrightarrow \check{H}^1_G(R_{\text{ét}_g}, F)$$
$$\longrightarrow \check{H}^1(R_{\text{ét}_g}, F) \longrightarrow \cdots$$

This is a tedious but straightforward computation. ∎

Proposition IV.1.10 allows us to construct a monomorphism $\mathrm{Br}_g(R) \to H^2_{\mathrm{gr}}(R_{\text{ét}_g}, U) \cong H^2(R_{\text{ét}_g}, U_0)$ in case R is quasistrongly graded. This approach was used by the first author in [10]. In fact, it is an application of the Knus-Ojanguren method (cf. [37]). However, it has some disadvantages; first, it will not work for general graded rings, because of the counterexample given at the end of Chapter II. Furthermore, it relies on the very deep refinement theorem of Artin.

A possible alternative is the use of Giraud's nonabelian cohomology (cf. [27]), as is done by Milne in the nongraded situation (cf. [44]). However, we prefer another approach, suggested to us by R. T. Hoobler, which is based on Verdier's contribution to [4] (exposé 5, sec. 7). Verdier replaces Čech cohomology by what he calls Čech cohomology, and uses another refinement theorem, which we will call *Verdier's refinement theorem.* It happens to work on any site, but we shall treat it only in our particular situation.

V.2 HYPERCOVERINGS AND VERDIER'S REFINEMENT THEOREM

V.2.1 We consider the category $\mathbf{\Delta}$ with objects $\Delta_n = [0,1,\ldots,n]$, for every integer n, and with nondecreasing maps for the morphisms. $\mathbf{\Delta}[n]$ is the full subcategory of $\mathbf{\Delta}$ with objects Δ_m, where $m \leq n$. By $\mathbf{\Delta} R_{\text{ét}_g}$, resp. $\mathbf{\Delta} R_{\text{ét}_g}[n]$, we denote the category of covariant functors from $\mathbf{\Delta}$, resp. $\mathbf{\Delta}[n]$, to $R_{\text{ét}_g}$. An obvious example of an object of $\mathbf{\Delta} R_{\text{ét}_g}$ is the Amitsur complex $1 \to S \to S^{(2)} \to \cdots$, for $S \in R_{\text{ét}_g}$, where the morphisms corresponding to the morphisms of $\mathbf{\Delta}$ are the maps obtained from the multiplication map $m : S^{(2)} \to S$, and from the canonical maps $\varepsilon_i : S^{(n)} \to S^{(n+1)}$, by composition. In general, an object of $\mathbf{\Delta} R_{\text{ét}_g}$ may be viewed as a sequence of objects of $R_{\text{ét}_g}$, namely, $1 \to T_0 \to T_1 \to \cdots$, together with maps $m : T_1 \to T_0$ and $\varepsilon_i : T_n \to T_{n+1}$ satisfying $m \circ \varepsilon_1 = m \circ \varepsilon_2 = 1, \cdots$. We shall refer to this object as $T_\bullet$. Now, consider the natural embedding $i_n : \mathbf{\Delta}[n] \to \mathbf{\Delta}$. Then i_n induces the truncation functor $i_n^* : \mathbf{\Delta} R_{\text{ét}_g} \to \mathbf{\Delta} R_{\text{ét}_g}[n]$.

V.2.2 Proposition i_n^* has a left adjoint functor $i_{n*} : \mathbf{\Delta} R_{\text{ét}_g}[n] \to \mathbf{\Delta} R_{\text{ét}_g}$. This means that for every object $F_\bullet = (0 \to F_0 \to F_1 \to \cdots \to F_n)$ in $\mathbf{\Delta} R_{\text{ét}_g}[n]$, and for every $G_\bullet = (0 \to G_0 \to G_1 \to \cdots)$ in $\mathbf{\Delta} R_{\text{ét}_g}$, we have a natural morphism

$$\mathrm{Hom}_{\mathbf{\Delta} R_{\text{ét}_g}[n]}(F_\bullet, i_n^* G_\bullet) \cong \mathrm{Hom}_{\mathbf{\Delta} R_{\text{ét}_g}}(i_{n*} F_\bullet, G_\bullet)$$

Proof: This is a very special case of [4, I.5.1]. We present the proof in our particular situation. Take an object Δ_p of $\mathbf{\Delta}$, and define the category I^p by the following objects and morphisms:
Objects: (Δ_i, f), where $i \leq n$, and where $f : \Delta_i \to \Delta_p$ is a nondecreasing map.
Morphisms: A morphism from (Δ_i, f) to (Δ_j, g) is a $\mathbf{\Delta}$-morphism $\xi : \Delta_i \to \Delta_j$ such that $f = g \cdot \xi$.

Let $h : \Delta_p \to \Delta_q$ be a $\mathbf{\Delta}$-morphism; by composition, we obtain a functor $I^h : I^p \to I^q$, and a commutative diagram

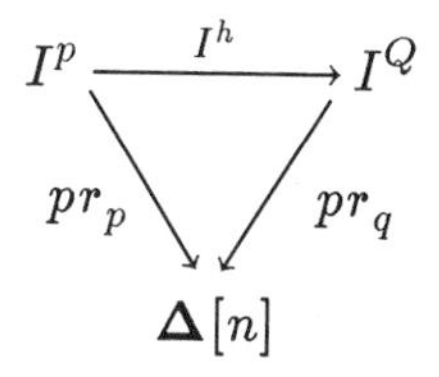

where $pr_p(\Delta_i, f) = \Delta_i$. For an object $F_\bullet$ of $\mathbf{\Delta} R_{\text{ét}_g}[n]$, we define $i_{n*}(F_\bullet) = K_\bullet$ by $K_p = \varinjlim(F \circ pr_p)$, where the inductive limit is taken over the finite category I^p. This limit exists in $\mathbf{\Delta} R_{\text{ét}_g}$, and $\mathbf{\Delta} R_{\text{ét}_g}$ has finite sums and amalgamated products (cf. [4, I 2.3]).

For a nondecreasing map $h : \Delta_p \to \Delta_q$, we have a functor $I^h : I^p \to I^q$ (see above), so we have a map $K_h : K_p \to K_q$, and we have therefore that $K_\bullet \in \mathbf{\Delta} R_{\text{ét}_g}$. Let us show that i_{n*} is a left adjoint to i_n^*. Take $\psi_\bullet \in \mathrm{Hom}_{\mathbf{\Delta} R_{\text{ét}_g}}(K_\bullet, G_\bullet)$. Then for all p, we have a morphism $\psi_p : K_p \to G_p$. Consider $(\Delta_p, 1_p)$ in the category I^p, where $p \leq n$. Then we have a canonical morphism $F_p \to K_p$, so, by composition, we obtain $\tilde{\psi}_p : K_p \to G_p$, and therefore a morphism $\tilde{\psi}_\bullet : F_\bullet \to i_n G_\bullet$ in $\mathbf{\Delta} R_{\text{ét}_g}[n]$.

Conversely, take $\tilde{\psi}_\bullet$ in $\mathbf{\Delta} R_{\text{ét}_g}[n]$. For every $i \leq n$, we have a morphism $\tilde{\psi}_i : F_i \to G_i$. Take $p \in \mathbf{N}$, and $(\Delta_i, f) \in I^p$. Then the map $f : \Delta_i \to \Delta_p$ induces $G_f : G_i \to G_p$, so we have a map $G_f \circ \tilde{\psi}_i : (F \circ pr_p)(\Delta_i, f) = F_i \to G_i \to G_p$, which behaves functorially. By definition of inductive limits, we have a canonical map $\psi_p : K_p = \varinjlim(F \circ pr_p) \to G_p$. This map defines the corresponding $\psi_\bullet : K_\bullet \to G_\bullet$. ∎

Note It follows easily from our construction that $F_p = K_p$ for $p \leq n$.

V.2.3 Examples (2.3.1) Let $n = 0$, and $F_0 = S$. Then $i_{0*}(F)$ is the Amitsur complex $0 \to S \to S^{(2)} \to \cdots$

(2.3.2) Let $n = 1$, and $F_0 = S$, $F_1 = T$. Denote $i_{1*}(F_\bullet) = K_\bullet$. Then, of course, $K_0 = S$ and $K_1 = T$. Now K_2 may be described as follows: consider the free abelian group with basis $\{a \boxtimes b \boxtimes c : a, b, c \in T\}$. Divide out the usual addition and multiplication relations for the tensor product, that is,

$$(a + a') \boxtimes b \boxtimes c = a \boxtimes b \boxtimes c + a' \boxtimes b \boxtimes c$$

$$(a \boxtimes b \boxtimes c)(a' \boxtimes b' \boxtimes c') = aa' \boxtimes bb' \boxtimes cc'$$

etc. Furthermore, divide out the following relations ($s \in S, a, b, c \in T$):

$$\varepsilon_1(s)a \boxtimes b \boxtimes c = a \boxtimes \varepsilon_1(s)b \boxtimes c$$

$$\varepsilon_2(s)a \boxtimes b \boxtimes c = a \boxtimes b \boxtimes \varepsilon_1(s)c$$

$$a \boxtimes \varepsilon_2(s)b \boxtimes c = a \boxtimes b \boxtimes \varepsilon_2(s)c$$

The reader may verify that, for $T = S^{(2)}$, we have $K_2 = S^{(3)}$. Then $K_3, K_4, \ldots$ may be constructed in a similar, but more complicated way.

V.2.4 The *coskeleton functor* $\mathrm{cosk}_n : \Delta R_{\text{ét}_g} \to \Delta R_{\text{ét}_g}$ is defined to be the composition $i_{n*} \circ i_n^*$. Observe that the canonical morphisms $i_n^* \circ i_{n*} \to Id$ and $\mathrm{cosk}_n \to \mathrm{cosk}_n \circ \mathrm{cosk}_m$ are isomorphisms for $m \leq n$ (cf. [4, I.7.1.1]).

V.2.5 Definition Take $p > 0$, and let $K_\bullet \in \Delta R_{\text{ét}_g}$. We call $K_\bullet$ a *graded étale hypercovering* of R if

(2.5.1) For all n, $(\mathrm{cosk}_n K_\bullet)_{n+1} \to K_{n+1}$ is a graded étale covering.

(2.5.2) $R \to K_0$ is a graded étale covering.

(2.5.3) If, furthermore, $K_\bullet \to \mathrm{cosk}_p(K_\bullet)$ is an isomorphism, then $K_\bullet$ is called a *hypercovering of type* p.

Let the following full subcategories of $\Delta R_{\text{ét}_g}$ be defined by their objects:

$\mathbf{H}R_p$: all hypercoverings of type p

$\mathbf{H}R_\infty$: all hypercoverings of finite type

$\mathbf{H}R$: all hypercoverings

V.2.6 Theorem (Verdier's Refinement Theorem) Let $K_\bullet \in \mathbf{H}R_p$ (resp. $\mathbf{H}R$), and take $0 \leq n \leq p$ (resp. $0 \leq n$). Suppose that T is a graded étale covering of K_n; then there exists $L_\bullet \in \mathbf{H}R_p$ (resp. $\mathbf{H}R$) such that we have a factorization $K_n \to T \to L_n$.

Proof: A general treatment would lead us too far, so we refer to Verdier [4, V.7.3.5]). Let us prove one special case that will be important in the sequel, namely, the case where $K_\bullet$ is the Amitsur complex. and where $n = 1$. Then we have the following situation:

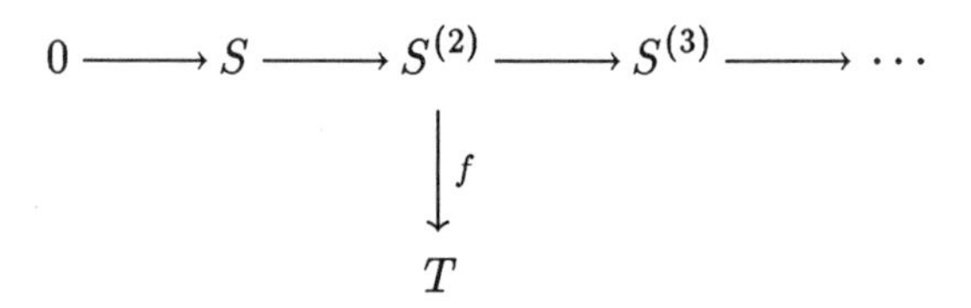

Let T_i be T, considered as an S-algebra using the morphism $f \circ \varepsilon_i$, and let $L_2 = T_1 \otimes_S T_2$. Then we have two maps ε_1, $\varepsilon_2 : L_2 \to L_2$, defined by $\varepsilon_1(a \otimes b) = ab \otimes 1$ and $\varepsilon_2(a \otimes b) = 1 \otimes ab$. Also note that we have a switch map on L_2. Consider the $S^{(2)}$-algebras $m : S^{(2)} \to S$ and $g : S^{(2)} \to L_2$, where $g = (f \circ \varepsilon_1) \otimes (f \circ \varepsilon_2)$, and define $L_1 = L_2 \otimes_2 S$. The following diagram then defines the map $m : L_2 \to L_1$ and $\varepsilon_i : L_1 \to L_2$ $(i = 1,2)$:

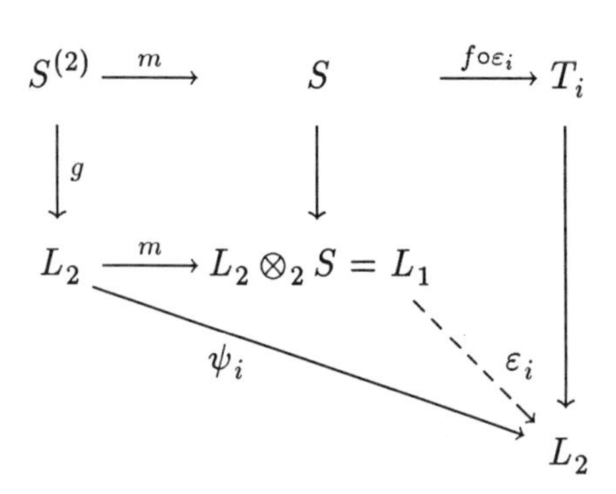

So $0 \to L_1 \to L_2$ is an object of $\mathbf{\Delta} R_{\text{ét}_g}[1]$. We define $L_\bullet = i_1(0 \to L_1 \to L_2)$. ∎

V.2.7 Let F, G be sheaves of abelian groups on $R_{\text{ét}_g}$, and $d : F \to G$. To an object $K_\bullet$ of $\mathbf{H}R$, we may associate a map of complexes, in a similar way as we did for the Amitsur complex in Chapter III:

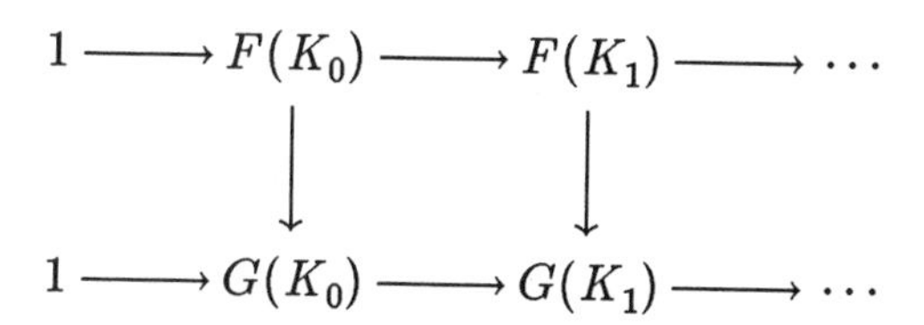

The associated cohomology groups are denoted $H^n(K_\bullet/R,F)$, $H^n(K_\bullet/R)$, $H^n_G(K_\bullet/R,F)$. An argument similar to the one in Chapter V in [37], or to V.7.3.2 in [4], shows that we may define the limit of these cohomology groups over all hypercoverings (of degree p). We denote these by $\check{H}^n(R_{\text{ét}_g},F)$; $\overset{p}{H}{}^n(R_{\text{ét}_g},F)$; $\check{H}^n_G(R_{\text{ét}_g},F)$; $\overset{p}{\check{H}}{}^n_G(R_{\text{ét}_g},F)$.

V.2.8 Theorem If $p > n-2$ holds, then $H^n_G(R_{\text{ét}_g},F) = \check{H}^n_G(R_{\text{ét}_g},F) = \overset{p}{\check{H}}{}^n_G(R_{\text{ét}_g},F)$, and $H^n(R_{\text{ét}_g},F) = \check{H}^n(R_{\text{ét}_g},F) = \overset{p}{\check{H}}{}^n(R_{\text{ét}_g},F)$.

Proof: A special case of [4, V.7.4.1]; the result follows from the lemmas below.

V.2.9 Lemma Let $0 \to E \to F \to G$ be an exact sequence in $\underline{S}(R_{\text{ét}_g})$; then we have an exact sequence $0 \to E(R) \to F(R) \to G(R) \to \check{H}^1_G(R_{\text{ét}_g},F) \to \check{H}^1(R_{\text{ét}_g},F) \to \check{H}^1(R_{\text{ét}_g},G) \to \cdots$. If $F \to G$ is an epimorphism, then $\check{H}^p_G(R_{\text{ét}_g},F) = \check{H}^p(R_{\text{ét}_g},E)$.

Proof: We leave this to the reader; one has to use V.2.6 and do some diagram chasing. ∎

V.2.10 Lemma Let F be an injective object of $\underline{S}(R_{\text{ét}_g})$. Then for all $n > 0$, and for each hypercovering $K_\bullet$ of R, $H^n(K_\bullet/R,F) = 0$.

Proof: We have to show that $0 \to F(K_0) \to F(K_1) \to \cdots$ is exact. Take $S \to T$ in $R_{\text{ét}_g}$, and consider the presheaf $\mathbf{Z}_T$ defined by letting $\mathbf{Z}_T(U)$ be the direct sum of $|\operatorname{Hom}_{R_{\text{ét}_g}}(T,U)|$ copies of $\mathbf{Z}$, where $\mathbf{Z}$ is the constant presheaf. Then for every presheaf P, $\operatorname{Hom}(\mathbf{Z}_T, P) = P(T)$. So we have to show that $0 \to \operatorname{Hom}(\mathbf{Z}_{K_0}, F) \to \operatorname{Hom}(\mathbf{Z}_{K_1}, F) \to \cdots$ is exact. Since F is a sheaf, $\operatorname{Hom}(\mathbf{Z}_{K_i}, F) = \operatorname{Hom}(a\mathbf{Z}_{K_i}, F)$. Since F is injective, it suffices to prove that $1 \leftarrow a\mathbf{Z}_{K_0} \leftarrow a\mathbf{Z}_{K_1} \leftarrow \cdots$ is exact in $R_{\text{ét}_g}$; this follows from [3, 8.14]. ■

V.3 APPLICATION TO THE GRADED BRAUER GROUP

As before, let R be a commutative graded ring, and $R \to S$ a graded étale covering. Recall from [37, II.8] that if F is a covariant functor from $R_{\text{ét}_g}$ to *Groups*, we may still define $H^1(S/R,F)$, which will be a set. Taking inductive limits, we obtain $\check{H}^1(R_{\text{ét}_g}, F)$. In the sequel, we will deal with the following sheaves of nonabelian groups on $R_{\text{ét}_g}$ ($n \in \mathbf{N}, \underline{d} \in \mathbf{Z}^n$):

$\mathrm{Gl}_n^0(\underline{d})$, given by $\mathrm{Gl}_n^0(\underline{d})(S) = \mathrm{Gl}_n^0(S)(\underline{d}) = \mathrm{Aut}_0(S^n(\underline{d}))$

$\mathrm{PGl}_n^0(\underline{d})$, given by

$$\mathrm{PGl}_n^0(\underline{d})(S) = \mathrm{PGl}_n^0(S)(\underline{d}) = \mathrm{Aut}_0(M_n(S)(\underline{d}))$$

Using the graded versions of the Morita theorems, we obtain an exact sequence of presheaves

$$0 \longrightarrow U_0 \longrightarrow \mathrm{Gl}_n^0(\underline{d}) \longrightarrow \mathrm{PGl}_n^0(\underline{d}) \longrightarrow \mathrm{Pic}_g$$

Taking associated sheaves, we obtain an exact sequence of sheaves:

$$0 \longrightarrow U_0 \longrightarrow \mathrm{Gl}_n^0(\underline{d}) \longrightarrow \mathrm{PGl}_n^0(\underline{d}) \longrightarrow a\,\mathrm{Pic}_g$$

V.3.1 Proposition There exist two canonical maps $\gamma_0 : \mathrm{PGl}_n^0(R)(\underline{d}) \to H^1_{\mathrm{gr}}(R_{\text{ét}_g}, U)$ and $\gamma_1 : \check{H}^1(R_{\text{ét}_g}, \mathrm{PGl}_n^0(\underline{d})) \to H^2_{\mathrm{gr}}(R_{\text{ét}_g}, U)$.

Proof: Let us first define γ_0. A graded automorphism $\varphi : M_n(R)(\underline{d}) \to M_n(R)(\underline{d})$ is induced by a graded isomorphism $f : R^n(\underline{d}) \to R^n(\underline{d}) \otimes I$, for some $I \in \underline{\underline{\mathrm{Pic}}}_g(R)$. Taking a gr-étale covering S of R such that $[I \otimes S] = 1$ in $\mathrm{Pic}^g(S)$, we obtain, by extension of scalars, a graded map $f' : S^n(\underline{d}) \to S^n(\underline{d}) \otimes T$, where $T \in \mathrm{gr}(S)$. Then f' defines a graded automorphism $g = f_2^{-1} f_1 : (S \otimes S)^n(\underline{d}) \otimes T_2 \to (S \otimes S)^n(\underline{d}) \otimes T_1 \otimes T_2 \to (S \otimes S)^n(\underline{d}) \otimes T_1$. Now g lies in the center of $M^n(S \otimes S)$, so it is given by multiplication by a unit $t \in U(S^{(2)})$, which is a cocycle. Clearly $d_1(t) = D_0(T)$, so we may define $\gamma_0(\varphi) = [(t,T)]$.

For the definition of γ_1, let us use the identification given in Theorem V.2.8. Let $\varphi : M_n(S \otimes S)(\underline{d}) \to M_n(S \otimes S)(\underline{d})$ represent an element of $\check{H}^1(R_{\text{ét}_g}, \mathrm{PGl}_n^0(\underline{d}))$. Then $\varphi_2 = \varphi_3 \varphi_1$, so φ is induced by a graded isomorphism $f : (S \otimes S)^n \otimes_2 I \to (S \otimes S)^n$, where $I \in \underline{\underline{\mathrm{Pic}}}_g(S^{(2)})$. Choose a graded étale covering T of $S^{(2)}$ such that $[I \otimes_2 T] = 1$ in $\mathrm{Pic}^g(S^{(2)})$. Using V.2.6., we obtain a hypercovering $L = (0 \to L_1 \to L_2 \to \cdots)$ such that $\varphi_{L_2} : M_n(L_2)(\underline{d}) \to M_n(L_2)(\underline{d})$ is induced by a graded map $f : L_2^n \otimes_{L_2} U \to L_2^n$, where U represents an element of $\mathrm{gr}(L_2)$. $f_2^{-1} f_3 f_1$ is multiplication by a unit u. We define $\gamma_1([\varphi]) = [(u,U)] \in \check{H}^2_{\mathrm{gr}}(R_{\text{ét}_g}, U) = H^2_{\mathrm{gr}}(R_{\text{ét}_g}, U)$. ■

Remark The maps γ_0, γ_1 defined above may be used to construct an exact sequence. For $S \in R_{\text{ét}}$, define the ordered set $\Gamma_S = \{(n,\underline{d}) : n \in \mathbf{N}_0, \underline{d} : \mathrm{Spec}(S) \to \mathbf{Z}^n \text{ continuous}\}$. Γ_S is ordered by divisibility, and if S is connected, then Γ_S is just the set Γ introduced in III.4.12. Let us define

$$\mathrm{Gl}^0(S) = \varinjlim \mathrm{Gl}_n^0(S)(\underline{d})$$

$$\mathrm{PGl}^0(S) = \varinjlim \mathrm{PGl}_n^0(S)(\underline{d})$$

where the inductive limits are taken over Γ_S. Then Gl^0 and PGl^0 define sheaves of nonabelian groups on $R_{\text{ét}_g}$. Let us define a map $\alpha_1 : H^1_{\mathrm{gr}}(R_{\text{ét}_g}, U) \to H^1(R_{\text{ét}_g}, \mathrm{Gl}^0)$. Let $x \in H^1_{\mathrm{gr}}(R_{\text{ét}_g}, U)$ be represented by (t,T), where $t \in U(S \otimes S)$, $T \in \mathrm{gr}(S)$, for some gr-étale covering $R \to S$. Multiplication by t induces a graded map $g : T_2 \to T_1$. Let $S \to S'$ be a gr-étale covering such that $T \otimes S' = T'$ is graded free on the connected components of S', that is, $T' = S'(\underline{d})$, where $\underline{d} : \mathrm{Spec}(S) \to \mathbf{Z}$ is continuous. Then we obtain a graded map $g' : (S' \otimes S')(\underline{d}) \to (S' \otimes S')(\underline{d})$, representing an element of $H^1(S'/R, \mathrm{Gl}_1^0(\underline{d}))$, hence defining an element of $\check{H}^1(R_{\text{ét}_g}, \mathrm{Gl}^0)$. The maps γ_0, α_1, γ_1 induce the following exact sequence:

$$1 \longrightarrow U_0(R) \longrightarrow \mathrm{Gl}^0(R) \longrightarrow \mathrm{PGl}^0(R) \longrightarrow H^1_{\mathrm{gr}}(R_{\text{ét}_g}, U)$$
$$\longrightarrow \check{H}^1(R_{\text{ét}_g}, \mathrm{Gl}^0) \longrightarrow \check{H}^1(R_{\text{ét}_g}, \mathrm{PGl}^0) \longrightarrow H^2_{\mathrm{gr}}(R_{\text{ét}_g}, U)$$

V.3.2 Lemma Let S be a graded étale covering of R. Then the set $H^1(S/R, \mathrm{PGl}_n^0(\underline{d}))$ classifies the set of graded isomorphism classes of graded R-Azumaya algebras A satisfying $A \otimes S \cong_g M_n(S)(\underline{d})$.

Proof: Consider a graded isomorphism $\sigma : A \otimes S \to M_n(S)(\underline{d})$. Then the map

$$\varphi = \sigma_2 \circ \tau_2 \circ \sigma_1^{-1} : S \otimes M_n(S)(\underline{d}) \longrightarrow S \otimes A \otimes S$$
$$\longrightarrow A \otimes S \otimes S \longrightarrow M_n(S)(\underline{d}) \otimes S$$

defines a graded descent datum of $M_n(S)(\underline{d})$, hence an element of $H^1(S/R, \mathrm{PGl}_n^0(\underline{d}))$. Conversely, take an element of $H^1(S/R, \mathrm{PGl}_n^0(\underline{d}))$, that is, a graded automorphism $\varphi : S \otimes M_n(S)(\underline{d}) \to M_n(S)(\underline{d}) \otimes S$ satisfying $\varphi_2 = \varphi_3 \circ \varphi_1$. Then φ is a graded descent datum for $M_n(S)(\underline{d})$, so it defines a graded R-Azumaya algebra A. ∎

V.3.3 Lemma For every graded R-Azumaya algebra A, there exists a graded étale covering S of R such that $[A \otimes S] = 1$ in $\mathrm{Br}_g(S)$.

Proof: Take $p \in \mathrm{Spec}^g(R)$. From Chapter IV, we know that there exists a graded Galois extension $S(p)$ of R_p^g splitting A_p^g in $\mathrm{Br}_g(R_p^g)$. So we can find $f \in h(R-p)$, and a graded Galois extension $S(f)$ of R_f splitting A_f. Clearly $\{D^g(f) : f$ associated to p for some $p \in \mathrm{Spec}^g(R)$ as above$\}$ covers $\mathrm{Spec}^g(R)$; hence there exists a finite subset $\{p_1, \ldots, p_n\}$ of $\mathrm{Spec}^g(R)$ such that $\{D^g(f_1), \ldots, D^g(f_n)\}$ covers $\mathrm{Spec}^g(R)$. Then $S(f_1) \times \cdots \times S(f_n)$ is a graded étale covering of R splitting A. ∎

From the preceding lemma, it follows that every graded Azumaya algebra of constant rank n^2 may be split by a graded étale covering S in such a way that $A \otimes S \cong_g M_n(S)(\underline{d})$. Indeed, we have a graded étale covering S such that $A \otimes S \cong_g \mathrm{END}_S(P)$. Take a graded Zariski covering S' of S such that $P \otimes_S S' \cong_g S'^n(\underline{d})$, and replace S by S'. We also remark that every graded Azumaya algebra is equivalent to one of constant rank.

Let $\mathrm{MBr}_g(R)$ be the group obtained from the set of graded isomorphism classes of graded Azumaya algebras by dividing out the relation $\sim$ defined by

$$A \sim B \Longleftrightarrow A \otimes M_n(R)(\underline{d}) \cong_g B \otimes M_m(R)(\underline{e})$$

for some $n, m \in \mathbf{N}$, $\underline{d} \in \mathbf{Z}^n$, $\underline{e} \in \mathbf{Z}^m$. Also, let $M_g(R)$ be the semigroup of graded isomorphism classes of graded faithfully projective R-modules, modulo the relation

$$P \sim Q \Longleftrightarrow P \otimes R^n(\underline{d}) \cong_g Q \otimes R^m(\underline{e})$$

From V.3.2, it follows that we have an injective mapping $H^1(R_{\text{ét}_g}, \mathrm{PGl}_n^0(\underline{d})) \to \mathrm{MBr}_g(R)$. In fact we have $\mathrm{MBr}_g(R) \cong \varinjlim H^1(R_{\text{ét}_g}, \mathrm{PGl}_n^0(\underline{d}))$, where the limit is taken over the index set Γ introduced in III.2.14. In a similar way, we have that $M_g(R) \cong \varinjlim H^1(R_{\text{ét}_g}, \mathrm{Gl}_n^0(\underline{d}))$. It follows from V.3.1 that we have an exact sequence

$$M_g(R) \longrightarrow \mathrm{MBr}_g(R) \longrightarrow H^2_{\mathrm{gr}}(R_{\text{ét}_g}, U)$$

Now let $\alpha \in \mathrm{Br}_g(R)$ be represented by A, where A is of constant rank. Then A defines an element of $\mathrm{MBr}_g(R)$. Let $i(\alpha)$ be the image of this element in $H^2_{\mathrm{gr}}(R_{\text{ét}_g}, U)$. Then $i(\alpha)$ is independent of the choice of α, since if $[A] = [B]$ in $\mathrm{Br}_g(R)$, then $A \otimes B^{\mathrm{opp}} \cong_g \mathrm{END}_R(P)$ lies in the image of $M_g(R)$. So we defined a map $i : \mathrm{Br}_g(R) \to H^2_{\mathrm{gr}}(R_{\text{ét}_g}, U)$, which is clearly an injection. We have shown:

V.3.4 Theorem Let R be a commutative graded ring; then we have a natural embedding $i : \mathrm{Br}_g(R) \to H^2_{\mathrm{gr}}(R_{\text{ét}_g}, U)$.

As a first application, let us study the injective behavior of $\mathrm{Br}_g(R) \to \mathrm{Br}(R)$.

V.3.5. Proposition If R is a commutative graded ring, and if $\mathrm{Pic}(R_p^{\mathrm{sgh}}) = 1$ for every graded prime ideal p of R, then
(3.5.1) $H^1(R_{\text{ét}_g}, U) \cong H^1(R_{\text{ét}}, U)$.
(3.5.2) $H^2(R_{\text{ét}_g}, U) \to H^2(R_{\text{ét}}, U)$ is a monomorphism.
(3.5.3) $\mathrm{Br}_g(R) \to \mathrm{Br}(R)$ is a monomorphism.
The condition on R_p^{sgh} is satisfied in case R is a graded Krull domain, or in case R is a trivially graded ring.

Proof: By V.1.5, we have a morphism of sites $F : R_{\text{ét}_g} \to R_{\text{ét}}$, inducing a functor $F_* : \underline{S}(R_{\text{ét}}) \to \underline{S}(R_{\text{ét}_g})$. Using V.1.4, we obtain a Leray spectral sequence

$$0 \longrightarrow H^1(R_{\text{ét}_g}, U) \longrightarrow H^1(R_{\text{ét}}, U) \longrightarrow R^1F_*U(R)$$
$$\longrightarrow H^2(R_{\text{ét}_g}, U) \longrightarrow H^2(R_{\text{ét}}, U)$$

According to [44, III.1.13], R^1F_*U is the sheaf associated to the following presheaf on $R_{\text{ét}_g} : S \to H^1(S_{\text{ét}}, U) = \mathrm{Pic}(S)$.

Now, for all $p \in \mathrm{Spec}^g(R)$, $\mathrm{Pic}(R_p^{\mathrm{sgh}}) = 1$; consequently $R^1F_*U = 1$, and (3.5.1–2) follow. To prove (3.5.3), observe first that we may suppose that R is reduced, by IV.3. Then consider the following diagram:

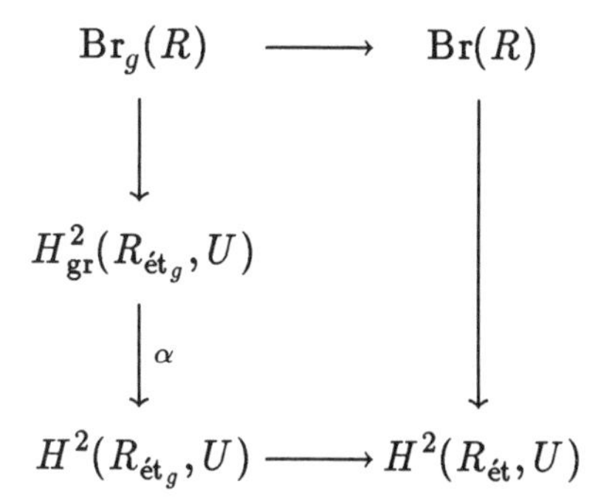

By V.3.4 and V.3.5.2, it is sufficient to show that α is a monomorphism. Consider the exact sequence $\cdots \to H^1(R_{\mathrm{\acute{e}t}_g}, \mathrm{gr}) \to H^2_{\mathrm{gr}}(R_{\mathrm{\acute{e}t}_g}, U) \to H^2(R_{\mathrm{\acute{e}t}_g}, U) \to \cdots$. As R is reduced, gr is just the constant sheaf $a\mathbf{Z}$ associated to the constant presheaf $\mathbf{Z}$. By [28, II.1.9], $H^1(R_{\mathrm{\acute{e}t}}, a\mathbf{Z}) = 0$. Furthermore, we have a Leray spectral sequence $1 \to H^1(R_{\mathrm{\acute{e}t}_g}, a\mathbf{Z}) \to H^1(R_{\mathrm{\acute{e}t}}, a\mathbf{Z}) \to R^1Fa\mathbf{Z}(R) \to \cdots$. Hence $H^1(R_{\mathrm{\acute{e}t}_g}, a\mathbf{Z}) = 1$ and the result follows.

Finally, for a graded Krull domain, $\mathrm{Pic}(R_p^{\mathrm{sgh}}) = \mathrm{Pic}^g(R_p^{\mathrm{sgh}}) = 1$, and for a trivially graded gr-local ring, R_p^{sgh} is trivially graded gr-local, hence local, such that its Picard group vanishes. ■

V.3.6 Corollary If R is a positively graded ring, then $\mathrm{Br}_g(R) \cong \mathrm{Br}(R_0)$.

Proof: By IV.3.1, $\mathrm{Br}_g(R) \cong \mathrm{Br}_g(R_0)$. The map $\mathrm{Br}_g(R_0) \to \mathrm{Br}(R_0)$ is injective by (3.5.3). As this map is clearly surjective, the result follows. ■

V.3.7 Another proof of (3.5.3) may be found in [15, 3.8]. A proof in the special case of a gr-Dedekind domain was given by the second author in [62]. Both proofs are different from the one presented above, and use techniques of a noncohomological nature.

A counterexample to (3.5.3) in the general situation was given by Van den Bergh in [58]. Before we give this example, let us state a lemma.

V.3.8 Lemma Suppose that

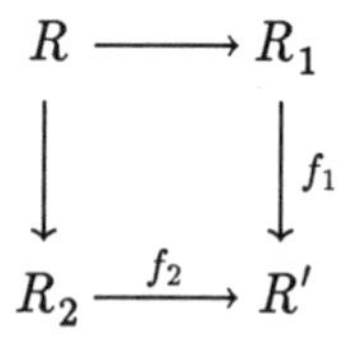

is a cartesian square of graded rings, where R' satisfies the following condition: if P is a graded projective R'-module of constant rank, then there exists graded free R'-modules F_0, F_1 such that $P \otimes F_0 \cong_g F_1$. Then we have a long exact sequence

$$\begin{aligned}1 &\longrightarrow U_0(R) \longrightarrow U_0(R_1) \oplus U_0(R_2) \longrightarrow U_0(R') \longrightarrow \mathrm{Pic}_g(R')\\ &\longrightarrow \mathrm{Pic}_g(R_1) \oplus \mathrm{Pic}_g(R_2) \longrightarrow \mathrm{Pic}_g(R') \longrightarrow \mathrm{Br}_g(R)\\ &\longrightarrow \mathrm{Br}_g(R_1) \oplus \mathrm{Br}_g(R_2) \longrightarrow \mathrm{Br}_g(R')\end{aligned}$$

Proof: Exactly as in [38]. Observe that the condition on R' is satisfied if R' is gr-semilocal and has nontrivially graded graded residue class fields (III.3.14). ∎

V.3.9 Example Let A be a one-dimensional noetherian reduced local ring with finite integral closure $\bar{A}$, and suppose that the maximal ideal of A splits in $\bar{A}$ into three primes. An example of such a ring is $k[[X,Y]]/(XY(X-Y))$. Let c be the conductor of $\bar{A}$ in A. Let $A' = A/c$, $\bar{A}' = \bar{A}/c\bar{A}$, $R = A[T,T^{-1}]$, $R' = A'[T,T^{-1}]$, $\bar{R} = \bar{A}[T,T^{-1}]$, $\bar{R}' = \bar{A}'[T,T^{-1}]$, and put $\deg T = 2$. We have

a cartesian square

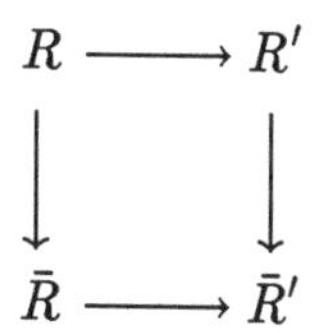

where $\bar{R}'$ satisfies the condition of the foregoing lemma. We therefore have the following diagram with exact top row:

$$\begin{array}{ccccc} \mathrm{Pic}_g(R') \oplus \mathrm{Pic}_g(\bar{R}) & \xrightarrow{\alpha} & \mathrm{Pic}_g(\bar{R}') & \xrightarrow{\beta} & \mathrm{Br}_g(R) \\ & & \downarrow & & \downarrow \gamma \\ & & \mathrm{Pic}\,\bar{R}') & \xrightarrow{\beta'} & \mathrm{Br}(R) \end{array}$$

The map β' is obtained using [38, th. 2.2]. As $\bar{A}'$ is Artinian, and $\bar{R}' = A'[T,T^{-1}]$, $\mathrm{Pic}(\bar{R}') = 1$, so $\gamma \circ \beta = 1$. In order to show that $\ker\gamma$ is nontrivial, it suffices therefore to show that $\operatorname{coker}\alpha$ is nontrivial. Consider the exact sequence

$$\begin{aligned} 1 &\longrightarrow U_0(R) \longrightarrow U(R) \longrightarrow \mathrm{gr}(R) \\ &\longrightarrow \mathrm{Pic}_g(R) \longrightarrow \mathrm{Pic}^g(R) \longrightarrow 1 \end{aligned}$$

Observe that $\mathrm{Pic}^g(R') = \mathrm{Pic}^g(\bar{R}) = \mathrm{Pic}^g(\bar{R}') = 1$, and $\mathrm{gr}(R') = \mathbf{Z}$, $\mathrm{gr}(\bar{R}) = \mathbf{Z}$, and $\mathrm{gr}(\bar{R}') = \mathbf{Z}^3$. So it follows that $\mathrm{Pic}_g(R') = \mathrm{Pic}_g(\bar{R}) = \mathbf{Z}/2\mathbf{Z}$ and $\mathrm{Pic}_g(\bar{R}') = \mathbf{Z}/2\mathbf{Z}^3$. Counting elements, we obtain that α can never be surjective.

V.3.10 Theorem Let I be a nilpotent graded ideal of R. Then for all $p \geq 1$ we have (denoting $\bar{R} = R/I$):
(3.10.1) $H^p(R_{\text{ét}_g}, U) \cong H^p(\bar{R}_{\text{ét}_g}, U)$,
(3.10.2) $H^p(R_{\text{ét}_g}, U_0) \cong H^p(\bar{R}_{\text{ét}_g}, U_0)$,
(3.10.3) $H^p(R_{\text{ét}_g}, \mathrm{gr}) \cong H^p(\bar{R}_{\text{ét}_g}, \mathrm{gr})$.

Proof: First, assume that $I^2 = 0$, and consider the following exact sequence on $R_{\text{ét}_g}$:

$$(3.10.4) \quad 0 \longrightarrow I \xrightarrow{e} U \longrightarrow i_* U \longrightarrow 1$$

where, for a graded étale extension $R \to S$, we have $I(S) = I \otimes S$, and $i_* U$ is the sheaf on $R_{\text{ét}_g}$ induced by $i : R \to R/I$. Then $iU(S) = U(S/IS)$. Let e be defined by $e(s)(x) = 1 + x$. The exactness then follows from the fact that $I^2 = 0$. We therefore have a long exact sequence

$$\cdots \longrightarrow H^p(R_{\text{ét}_g}, I) \longrightarrow H^p(R_{\text{ét}_g}, U) \longrightarrow H^p(R_{\text{ét}_g}, i_* U)$$
$$\cong H^p(\bar{R}, U) \longrightarrow H^{p+1}(R_{\text{ét}_g}, U) \longrightarrow \cdots$$

The assertion now follows from the fact that $H^p(R_{\text{ét}_g}, I) = 0$.

Next, suppose that $I^n = 0$, let r be an integer greater than or equal to $n/2$, and let $\{x_1, \ldots, x_r : x_i \in I\}$ generate the graded ideal J. Then $J^2 = 0$. We apply the case $n = 2$ to J, and proceed by induction. This finishes the proof of (3.10.1). Then (3.10.2) is done in a similar way, but now using the exact sequence

$$0 \longrightarrow I_0 \longrightarrow U_0 \longrightarrow i_* U_0 \longrightarrow 1$$

From IV.3.7, it follows that the sheaves $a\,\text{Pic}_g$ and $i_* a\,\text{Pic}_g$ coincide on $R_{\text{ét}_g}$. So, in the case $I^2 = 0$, we have the following diagram with exact rows and columns:

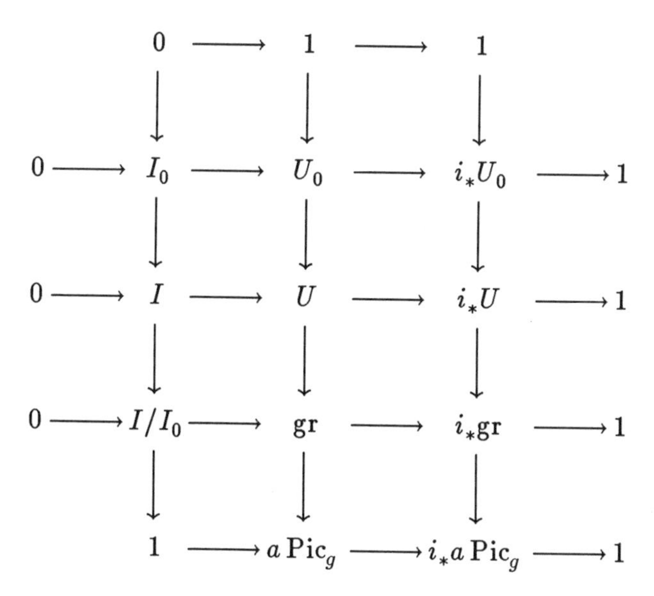

A diagram chase yield that indeed I/I_0 is the kernel of the map $\mathrm{gr} \to i_*\mathrm{gr}$. The proof of (3.10.3) is then finished in a similar way as the proof of (3.10.1–2). ∎

V.3.11 Corollary Let R be a noetherian graded ring. Then
(3.11.1) $H^1(R_{\text{ét}_g}, \mathrm{gr}) = 1$.
(3.11.2) $H^2_{\mathrm{gr}}(R_{\text{ét}_g}, U) \to H^2_{\mathrm{gr}}(R_{\text{ét}_g}, U)$ is monomorphic.
(3.11.3) $\mathrm{Pic}^g(R) \cong H^1(R_{\text{ét}_g}, U)$.
(3.11.4) $\mathrm{Pic}_g(R) \cong H^1_{\mathrm{gr}}(R_{\text{ét}_g}, U) (\cong H^1(R_{\text{ét}_g}, U_0)$ if R is quasistrongly graded).

Proof: (3.11.1) follows from (3.10.3) and the proof of V.3.5; (3.11.2) follows from (3.11.1) and the exact sequence (1.1.2); (3.11.3) follows from (3.11.1) and III.5.7. Finally, consider the exact sequence $1 \to U_0(R) \to U(R) \to \mathrm{gr}(R) \to H^1_{\mathrm{gr}}(R_{\text{ét}_g}, U) \to H^1(R_{\text{ét}_g}, U) \to H^1(R_{\text{ét}_g}, \mathrm{gr}) = 1$. Apply the five-lemma to this sequence and the sequence III.3.4, taking (3.11.3) into account. Then (3.11.4) follows. ∎

V.3.12 Theorem $\mathrm{Br}_g(R)$ is a torsion group.

Proof: Let us use the (small) graded flat site, as introduced in V.1. Note that there is no problem in constructing a map $\check{H}^1(R_{\mathrm{fl}_g}, \mathrm{PGl}^0_n(\underline{d})) \to H^2_{\mathrm{gr}}(R_{\mathrm{fl}_g}, U)$, as in V.3.1, and a monomorphism $\mathrm{Br}_g(R) \to H^2_{\mathrm{gr}}(R_{\mathrm{fl}_g}, U)$, as in V.3.4. Consider the sheaves μ^0_n and μ_n on R_{fl_g} defined by

$$\mu_n(S) = \{x \in S : x^n = 1\}$$

$$\mu^0_n(S) = \{x \in S_0 : x^n = 1\}$$

Then the Kümmer sequence

$$1 \longrightarrow \mu^0_n \longrightarrow U_0 \longrightarrow U_0 \longrightarrow 1$$

is exact on R_{fl_g} (not on $R_{\mathrm{\acute{e}t}_g}$ in general). The map $\mu_0 \to U_0$ is obtained by taking nth powers. Also, we have an exact sequence of sheaves on R_{fl_g}

$$1 \longrightarrow \mu^0_n \longrightarrow \mu_n \longrightarrow \mathrm{gr}$$

so we may define the cohomology groups $H^n_{\mathrm{gr}}(R_{\mathrm{fl}_g}, \mu_n)$. As in V.3.1, we may define a map $\delta_1 : \check{H}^1(R_{\mathrm{fl}_g}, \mathrm{PGl}^0_n(\underline{d})) \to H^2_{\mathrm{gr}}(R_{\mathrm{fl}_g}, \mu_n)$. Indeed, take a graded automorphism φ of $M_n(S \otimes S)(\underline{d})$ satisfying $\varphi_2 = \varphi_3 \circ \varphi_1$. In V.3.1, we defined a graded isomorphism $f : L^n_2 \otimes_{L_2} U \to L^n_2$, where U represents an element of $\mathrm{gr}(L_2)$. Now, if we suppose that S contains nth roots of unity, then we can construct f in such a way that f has determinant one. Now, as observed in V.3.1, $f_2^{-1} f_3 f_1$ is multiplication by a unit u, so u has to be an nth root of unity. Thus we may define $\delta_1([\varphi]) = [(u, U)]$. We clearly have a map $H^2_{\mathrm{gr}}(R_{\mathrm{fl}_g}, \mu_n) \to H^2_{\mathrm{gr}}(R_{\mathrm{fl}_g}, U)$, so we obtain the following commutative diagram:

$$\begin{array}{ccc}
\check{H}^1(R_{\mathrm{fl}_g}, \mathrm{PGl}_n^0(\underline{d})) & \longrightarrow & H^2_{\mathrm{gr}}(R_{\mathrm{fl}_g}, \mu_n) \\
\downarrow & & \downarrow \gamma \\
\check{H}^1(R_{\mathrm{fl}_g}, \mathrm{PGl}_n^0(\underline{d})) & \longrightarrow & H^2_{\mathrm{gr}}(R_{\mathrm{fl}_g}, U)
\end{array}$$

So γ factors through $H^2_{\mathrm{gr}}(R_{\mathrm{fl}_g}, \mu_n)$. Now $H^2_{\mathrm{gr}}(R_{\mathrm{fl}_g}, \mu_n)$ maps injectively into $H^2(R_{\mathrm{fl}_g}, \mu_n)$, which is n-torsion. Hence the image of $H^1(R_{\mathrm{fl}_g}, \mathrm{PGl}_n^0(\underline{d}))$ is n-torsion. Hence $\mathrm{Br}_g(R)$ is torsion, at least when R is connected. The general case follows easily. ∎

V.3.13 Note If R is quasistrongly graded, then V.3.12 may be proved very easily: invoking II.4.17, we can adapt the easy part of the Knus-Ojanguren proof [37, IV.6.1]. However, the general part of this proof, which avoids the use of Artin's theorem, does not admit a graded counterpart.

V.3.14 Note Consider the natural functor $i : \underline{S}(R_{\text{ét}_g}) \to \underline{P}(R_{\text{ét}_g})$, and let S be a gr-étale covering of R. We have a functor $g = H^0(S/R, \cdot) : \underline{P}(R_{\text{ét}_g}) \to \underline{Ab}$. It is easily seen that i takes injective sheaves to g-acyclic presheaves, so we may write down the Leray spectral sequence V.1.4. As $g \circ i = \Gamma$, we have that $R^n(g \circ i)(F) = H^n(R_{\text{ét}_g}, F)$. Also, $R^n g(P) = H^n(S/R, P)$ for every presheaf P, and $R^n f$ is the functor mapping a sheaf F to the presheaf $H^n(\cdot_{\text{ét}_g}, F)$ on $R_{\text{ét}_g}$; so if we take $F = U_0$, the sequences becomes

$$\begin{aligned}
0 &\longrightarrow H^1(S/R, U_0) \longrightarrow H^1(R_{\text{ét}_g}, U_0) \\
&\longrightarrow H^0(S/R, H^1(\cdot_{\text{ét}_g}, U_0)) \longrightarrow H^2(S/R, U_0) \\
&\longrightarrow \mathrm{Ker}(H^2(R_{\text{ét}_g}, U_0) \longrightarrow H^0(S/R, H^2(\cdot_{\text{ét}_g}, U_0)))
\end{aligned}$$

$$= \mathrm{Ker}(H^2(R_{\text{ét}_g}, U_0) \longrightarrow H^2(S_{\text{ét}_g}, U_0))$$
$$\longrightarrow H^1(S/R, H^1(\cdot_{\text{ét}_g}, U_0)) \longrightarrow H^3(S/R, U).$$

Now, if we assume that R is strongly graded on the gr-étale site, then we may prove directly, using descent theory, that $H^0(\cdot_{\text{ét}_g}, U_0) = \mathrm{Pic}_g$. If furthermore S is faithfully projective as an R-module, we have that $\mathrm{Ker}(H^2(R_{\text{ét}_g}, U_0) \to H^2(S_{\text{ét}_g}, U_0)) = \mathrm{Br}_g(S/R)$, and the sequence above becomes the Chase–Rosenberg sequence III.5.5.

V.4 A GRADED VERSION OF GABBER'S THEOREM

Let R be a ring which is strongly graded on the graded étale site. Recall (V.3) that we have a monomorphism $\mathrm{Br}_g(R) \to H^2(R_{\text{ét}_g}, U_0) \cong H^2_{\mathrm{gr}}(R_{\text{ét}_g}, U) \cong \check{H}^2(R_{\text{ét}_g}, U_0)$. In this section we shall show that $\mathrm{Br}_g(R) \cong \check{H}^2(R_{\text{ét}_g}, U_0)_{\mathrm{tors}}$. We use the approach of Knus and Ojanguren to this kind of problem (cf. [39]). For other proofs of the corresponding ungraded case, we refer to Gabber [26] and Hoobler [31]. First, we need a generalization of the Mayer–Vietoris sequence V.3.8. Consider a diagram of graded ring homomorphisms $R_1 \to R \leftarrow R_2$. We define a Brauer group relative to this diagram: consider data of the form (A_1, A_2, P, Q, ξ), where $A_i \in \underline{\underline{Az}}_g(R_i), P, Q \in \underline{\underline{FP}}_g(R)$, and where ξ is a graded isomorphism

$$\xi : A_1 \otimes_{R_1} R \otimes_R \mathrm{END}_R(P) \longrightarrow A_2 \otimes_{R_2} R \otimes_R \mathrm{END}_R(Q)$$

This implies that A_1 and A_2 become equivalent in $\mathrm{Az}_g(R)$. A datum of the form $(\mathrm{END}_R(P_1), \mathrm{END}_R(P_2), N, M, \mathrm{END}(\varphi))$, where φ is induced by a graded isomorphism $\varphi : P_1 \otimes_{R_2} R \otimes_R N \to P_2 \otimes_{R_2} R \otimes_R M$ is then called a *trivial* datum. Defining tensor products of data in the obvious way, two data Δ and Δ' are called *equivalent* if there exist trivial data E and E' such that $\Delta \otimes E \cong_g \Delta' \otimes$

E'. The equivalence classes of data then form a group, denoted $\mathrm{Br}_g(R_1, R_2, R)$.

V.4.1 Theorem Let R be a commutative graded ring that is strongly graded on the gr-étale site, and let f, $g \in h(R)$ be such that $Rf + Rg = R$. Then there exists a monomorphism $\eta : \mathrm{Br}_g(R_f, R_g, R) \to H^2(R_{\text{ét}_g}, U_0)$ and a commutative diagram of exact sequences

$$1 \longrightarrow U_0(R) \longrightarrow U_0(R_f) \oplus U_0(R_g) \longrightarrow U_0(R_{fg})$$
$$\longrightarrow \mathrm{Pic}_g(R) \longrightarrow \mathrm{Pic}_g(R_f) \oplus \mathrm{Pic}_g(R_g) \longrightarrow \mathrm{Pic}_g(R_{fg})$$

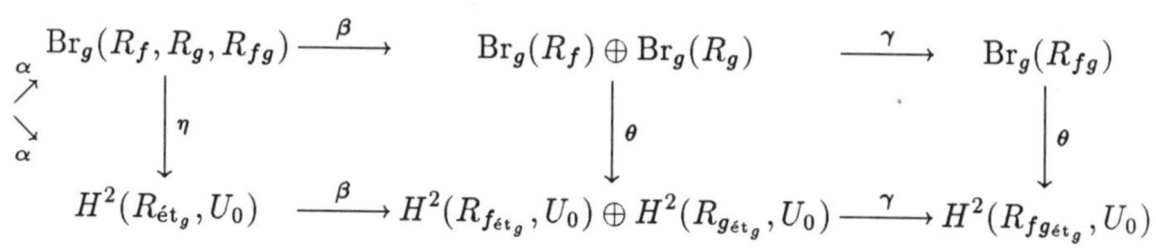

Proof: For details on the construction of the connecting maps, we refer to [39]. The proof is identical, up to the observation that all units and homomorphisms involved are of degree zero. Note that this proof relies on Artin's refinement theorem II.4.17. ∎

V.4.2 Lemma With notation and assumptions as in V.4.1, we have that, if $c \in \check{H}^2(R_{\text{ét}_g}, U_0)_{\text{tors}}$ is such that $\beta(c) \in \mathrm{Im}\,\theta$, then $c \in \mathrm{Im}(\mathrm{Br}_g(R) \to \check{H}^2(R_{\text{ét}_g}, U_0))$.

Proof: There exists $A \in \underline{\underline{Az}}_g(R_f)$, $B \in \underline{\underline{Az}}_g(R_g)$ such that $\theta([A]) = c_f$, $\theta([B]) = c_g$. Hence $\theta([A_g]) = \theta([B_f]) = c_{fg}$, so there exists a datum of the form $\Delta = (A, B, P, Q, \xi)$ such that $\theta \circ \beta([\Delta]) = (c_f, c_g)$. Replacing Δ by $\Delta \otimes (R_f, R_g, R^{\dagger}_{fg}, R^{\dagger}_{fg}, id)$ (using the notation of III.2 and III.3), we may suppose $\Delta = (A, B, P^{\dagger}, Q^{\dagger}, \xi)$. Again, replacing Δ by $\Delta \otimes (R_f, R_g, R_{fg}, I, \mathrm{can})$ for some suitable $I \in \underline{\underline{\mathrm{Pic}}}_g(R_{fg})$, we may suppose that $\eta([\Delta]) = c$. Now assume that $c^n = 1$; then, by injectivity of η, $\Delta^{(n)}$ is graded isomorphic to a trivial datum $(\mathrm{END}\,M, \mathrm{END}\,N, P^{\dagger(n)}, Q^{\dagger(n)}, \mathrm{END}\,\omega)$, where ω is a graded isomorphism $M_g \otimes P^{\dagger(n)} \to N_f \otimes Q^{\dagger(n)}$. Let $x = [M_g^{\dagger} \otimes P^{\dagger(n)}] = [N_f^{\dagger} \otimes Q^{\dagger(n)}]$ in $K_0\,\underline{\underline{FP}}_g^{\dagger}(R_{fg})$. With notation as in III.3.24, we let $y = x/\sigma(rk_g(x)) = \rho(x)$, and let z be the unique nth root of y^{-1} (it is clear that $rk_g(y) = 1$). Then $(z\rho([P^{\dagger}]))^n = \rho([M_g^{\dagger}])^{-1}$, and therefore $z\rho[P^{\dagger}] \in \mathrm{Im}(K_0\,\underline{\underline{FP}}_g^{\dagger}(R_f) \to K_0\,\underline{\underline{FP}}_g^{\dagger}(R_{fg}))$, and a similar property holds for $z\rho([Q^{\dagger}])$.

Consider the isomorphism $h : K_0\,\underline{\underline{FP}}_g^{\dagger}(R_{fg}) \to U^{+}(Q \otimes K_{og}^{\dagger}(R_{fg}))$. There exists an integer m such that $h([R^{\dagger})^m]z) = 1 \otimes \alpha$, for some α in $K_{og}^{\dagger}(R_{fg})$. Taking m big enough at once, we can arrange that $\alpha = [H^{\dagger}]$, using III.3.13. Hence $[(R^{\dagger})^m]z = [H^{\dagger}]$, for some $H^{\dagger} \in \underline{\underline{FP}}_g^{\dagger}(R_{fg})$, and, in a similar way,

$$[(R^{\dagger})^m]z[P^{\dagger}] = [P_g'^{\dagger}] \qquad \text{for some } P'^{\dagger} \in \underline{\underline{FP}}_g^{\dagger}(R_f)$$
$$[(R^{\dagger})^m]z[Q^{\dagger}] = [Q_f'^{\dagger}] \qquad \text{for some } Q'^{\dagger} \in \underline{\underline{FP}}_g^{\dagger}(R_g)$$

It follows that $[H^{\dagger}][P^{\dagger}] = [P_g'^{\dagger}]$, and therefore $H^{\dagger} \otimes P^{\dagger} \otimes (R^{\dagger})^p \cong_g P_g'^{\dagger} \otimes (R^{\dagger})^p$, for some integer p. Replacing m by mp, we obtain $H^{\dagger} \otimes P^{\dagger} \cong_g P_{g'}'^{\dagger}$ and similarly, $H^{\dagger} \otimes Q^{\dagger} \cong_g Q_g'^{\dagger}$. Now $\Delta \sim (A, B, P^{\dagger}, Q^{\dagger}, \xi) \otimes (R_f, R_g, H^{\dagger}, H^{\dagger}, Id) \cong_g (A, B, (P'^{\dagger})_g, (Q'^{\dagger})_f, \xi')$. Also, $A \otimes \mathrm{END}(P'^{\dagger})$ and $B \otimes \mathrm{END}(Q'^{\dagger})$ may be glued together using ξ'. This gives a graded R-Azumaya algebra C, and from our construction it follows that $\eta([C]) = c$. ∎

V.4.3 Lemma With notation as above, let $c \in \check{H}^2(R_{\text{ét}_g}, U_0)$, and denote $\beta'(c) = (c_f, c_g)$. If $Q_m^g(c) \in \mathrm{Im}(\mathrm{Br}_g(Q_m^g(R)) \to \check{H}^2(Q_m^g(R)_{\text{ét}_g}, U_0))$ for every m in $\mathrm{Max}_g(R)$, then we have that c is in $\mathrm{Im}(\mathrm{Br}_g(R) \to \check{H}^2(R_{\text{ét}_g}, U_0))$.

Proof: Take $m \in \mathrm{Max}^g(R)$. Then there exists $f \in h(R \setminus m)$, and a graded R_f-Azumaya algebra $A(f)$ such that $\theta([A(f)]) = c_f$. Consider $\Sigma = \{f \in h(R) : c_f \in \mathrm{Im}\,\theta\}$. It is easily seen that

(i) $a \in h(R), f \in \Sigma \Rightarrow af \in \Sigma$.

(ii) $f, g \in \Sigma, \deg f = \deg g \Rightarrow f + g \in \Sigma$ (applying V.4.2 to R_{f+g}).

Now $\mathrm{Spec}^g(R)$ can be covered by sets of the form $\{D^g(f(m)) : m \in \max^g(R), f(m) \in \Sigma\}$. By quasicompactness, this may be reduced to a finite covering $\{D^g(f_i) : i = 1, \ldots, n, f_i \in \Sigma\}$. Hence there exist homogeneous $a_1, \ldots, a_n$ such that $\Sigma_{i=1}^n a_i f_i = 1$. Hence $1 \in \Sigma$, finishing the proof. ∎

V.4.4 Lemma If $R = R_0[X, X^{-1}]$, with $\deg X = 1$, then $\mathrm{Br}_g(R) \cong H^2(R_{\text{ét}_g}, U_0)_{\text{tors}}$.

Proof: In this situation, $\mathrm{Br}_g(R) \cong \mathrm{Br}(R_0)$, since R is strongly graded. If S is a graded étale covering of R, then $S = S_0[X, X^{-1}]$, where S_0 is an étale covering of R_0. Clearly $S^{(n)} = S_0^{(n)}[X, X^{-1}]$; hence $U_0(S^{(n)}) = U(S_0^{(n)})$, and $H^n(S/R, U_0) = H^n(S_0/R_0, U)$. The result follows therefore from the classical theorem. ∎

V.4.5 Lemma Let R be a gr-local ring which is strongly graded on the graded étale site. Then $\mathrm{Br}_g(R) \cong \check{H}^2(R_{\text{ét}_g}, U_0)_{\text{tors}}$.

Proof: If R is gr-local and quasistrongly graded, then R has a homogeneous invertible element of positive degree d. Consider $C = R[X]/(X^d - T)$. Then C is a graded étale covering of R which is finitely generated as an R-module. Hence C is projective as an R-module, so C is graded free. The result now follows from the foregoing and the following lemma. ∎

V.4.6 Lemma Let C/R be a graded, faithfully projective R-algebra. Let $c \in \check{H}^2(R_{\text{ét}_g}, U_0)$ be such that its image in $\check{H}^2(C_{\text{ét}_g}, U_0)$ lies in $\mathrm{Im}\,\theta$. Then c itself lies in $\mathrm{Im}\,\theta$.

Proof: By V.4.3, we may suppose that R is gr-local. Let $R \to S$ be a graded étale covering, and let $u \in U_0(S^{(3)})$ represent c.

Let A be a graded C-Azumaya algebra such that $\theta([A]) = 1 \otimes c \in \check{H}^2(C_{\text{ét}_g}, U_0)$. As C is a finite graded extension of R, every graded étale covering of $C \otimes R^{\text{sgh}}$ has a section (see II.3.20). Hence $A \otimes R^{\text{sgh}}$ represents the trivial element of $\text{Br}_g(C \otimes R^{\text{sgh}})$, and thus there exists a graded étale covering T of R such that $[A \otimes T] = 1$ in $\text{Br}_g(C \otimes T)$. So we can find a graded isomorphism $\alpha : A \otimes T \to M_n(C \otimes T)$, if we suppose that A is of constant rank (which is no problem) and we replace T by another graded étale covering. Replacing T by yet another graded étale covering, and invoking II.4.17, we may suppose that the automorphism $(\alpha \otimes 1) \circ \tau \circ (1 \otimes \alpha)^{-1}$ of $M_n(C \otimes T \otimes T)$ is induced by a graded automorphism $D(\alpha)$ of $((C \otimes T) \otimes_C (C \otimes T))^n$. Replace S and T by the graded covering R^{sgh}. The cocycle f induced by $D(\alpha)$ is cohomologous to $1 \otimes u$; hence there exists a graded étale covering W of C, and $w \in U_0(W \otimes_C W)$, such that $1 \otimes u = f \Delta_1 w$. As $C \otimes R^{\text{sgh}} \to W \otimes R^{\text{sgh}}$ has a section, we may replace W first by $W \otimes R^{\text{sgh}}$ and then by $C \otimes R^{\text{sgh}}$. So we may choose w in $U_0((C \otimes R^{\text{sgh}}) \otimes_C (C \otimes R^{\text{sgh}}))$. Now, the graded isomorphism

$$D(\alpha)w : (C \otimes R^{\text{sgh}} \otimes R^{\text{sgh}})^n \longrightarrow (C \otimes R^{\text{sgh}} \otimes R^{\text{sgh}})^n$$

is an isomorphism of graded free $R^{\text{sgh}} \otimes R^{\text{sgh}}$-modules of rank nd, where $[C : R] = d$. So it induces a graded isomorphism of graded $R^{\text{sgh}} \otimes R^{\text{sgh}}$-algebras,

$$\begin{aligned} \psi = \text{END}(D(\alpha)w) &: M_{nd}(R^{\text{sgh}} \otimes R^{\text{sgh}}) \\ &\longrightarrow M_{nd}(R^{\text{sgh}} \otimes R^{\text{sgh}}) \end{aligned}$$

Clearly we have $\psi_2 = \psi_3 \psi_1$, so by descent we obtain a graded Azumaya algebra B. By construction, the cocycle associated to it is u. ■

V.4.7 Theorem (Graded Version of Gabber's Theorem) If R is strongly graded on the gr-étale site, then $\text{Br}_g(R) \cong H^2(R_{\text{ét}_g}, U_0)_{\text{tors}}$.

Proof: From V.4.3 and V.4.6. ∎

V.4.8 Note The following question is open at present: Assume that R is strongly graded on the gr-flat site. Do we have an isomorphism $\mathrm{Br}_g(R) \cong H^2_{\mathrm{gr}}(R_{\text{ét}_g}, U)_{\mathrm{tors}}$?

We now give an application of the techniques used in the proof of V.4.7. Let R be a Krull domain, and let $R_{(n)}$ be the ring obtained by blowing up the grading of R by a factor n. We have a monomorphism $\mathrm{Br}_g(R) \to \mathrm{Br}(R)$. We shall show below that $U\mathrm{Br}_g(R) = \bigcup_{n>0} \mathrm{Br}_g(R_{(n)}) = \mathrm{Br}(R)$, if R is quasistrongly graded and contains a field of characteristic zero, thus generalizing IV.1.11 and IV.2.8.

V.4.9 Lemma Let R be strongly graded on the gr-étale site, and let f, g be as in V.4.1. Consider the monomorphisms i : $\mathrm{Br}_g(R) \to \mathrm{Br}_g(R_f, R_g, R_{fg})$ and i' : $\mathrm{Br}(R) \to \mathrm{Br}(R_f, R_g, R_{fg})$. Then $\mathrm{Im}\, i = \mathrm{Br}_g(R_f, R_g, R_{fg})_{\mathrm{tors}}$ and $\mathrm{im}\, i' = \mathrm{Br}(R_f, R_g, R_{fg})_{\mathrm{tors}}$.

Proof: From V.4.1 and V.4.7 and [39, 3.1 and 3.2]. ∎

V.4.10 Theorem Let R be a Krull domain which is strongly graded on the étale site, and suppose that $n^{-1} \in R$. Then ${}_n\mathrm{Br}(R) = {}_n\mathrm{Br}_g(R_{(n)})$.

Proof: Denote $T = R_{(n)}$. By IV.2.8, we know that $\mathrm{Br}(Q^g_p(R)) = {}_n\mathrm{Br}_g(Q^g_p(T))$ for every $p \in \mathrm{Spec}_g(R)$. Let f, g be as in V.4.1, and consider the following diagram with exact rows:

$$\begin{array}{ccccccc}
\mathrm{Pic}^g(T_{fg}) & \xrightarrow{\alpha} & \mathrm{Br}_g(T_f, T_g, T_{fg}) & \xrightarrow{\beta} & \mathrm{Br}_g(T_f) \oplus \mathrm{Br}_g(T_g) & \xrightarrow{\gamma} & \mathrm{Br}_g(T_{fg}) \\
\downarrow \cong & & \downarrow \psi & & \downarrow \varphi & & \downarrow \rho \\
\mathrm{Pic}(T_{fg}) & \xrightarrow{\alpha'} & \mathrm{Br}(T_f, T_g, T_{fg}) & \xrightarrow{\beta'} & \mathrm{Br}(T_f) \oplus \mathrm{Br}(T_g) & \xrightarrow{\gamma'} & \mathrm{Br}(T_{fg})
\end{array}$$

It is left to the reader to show that, in this particular situation, we may write $\mathrm{Pic}^g(T_{fg})$ instead of $\mathrm{Pic}_g(T_{fg})$ in the top row. Take $A \in {}_n\mathrm{Br}(T)$, and suppose that $\beta' \circ i'([A]) \in \mathrm{Im}(\varphi)$. Some diagram chasing yields that $i'([A]) \in \mathrm{Im}(\psi)$, so $i'([A]) = \psi([\Delta])$ for some datum Δ. As $i'([A])$ is n-torsion in $\mathrm{Br}(T_f, T_g, T_{fg})$, $[\Delta]$ is n-torsion, so $[\Delta] \in {}_n\mathrm{Br}_g(T_f, T_g, T_{fg}) = \mathrm{Im}(i)$. Hence $[A] \in \mathrm{Im}({}_n\mathrm{Br}_g(T) \to \mathrm{Br}(T))$. The proof is now finished by using an induction argument as in V.4.3. Note that we used the fact that $\mathrm{Br}_g(R) \to \mathrm{Br}(R)$ is into (R is a Krull domain!). ■

V.5 THE VILLAMAYOR–ZELINSKY APPROACH

This section contains no new results, but gives an alternative approach to the Chase–Rosenberg sequences and to the embedding $\mathrm{Br}_g(R) \to H^2_{\mathrm{gr}}(R_{\text{ét}_g}, U)$. The techniques involved are based on the work of Villamayor and Zelinsky (cf. [72, 74]). We include it as a separate section, as it gives a proof of the embedding, without using a refinement theorem. Actually, all the applications of Section V.3 follow from it, except for the result concerning the torsion of the graded Brauer group. Furthermore, it is shown that the first of the graded Chase–Rosenberg sequences III.5.5 appears as a Leray spectral sequence, at least in the case where R is strongly graded on the graded étale site.

The link between the second Chase–Rosenberg sequence III.5.8 and the Leray spectral sequences of U and gr, however, is less obvious. Throughout this section, we assume that R is noetherian. The graded étale site may be replaced by the graded flat site in all applications below.

V.5.1 Let R be a commutative graded ring, and S a commutative faithfully flat graded R-algebra. Following [74, 75], we define abelian groups E_n, for $n > 0$, as follows. Take $I \in \underline{\underline{\mathrm{Pic}}}_g(S^{(n)})$, and define $\delta_{n-1}I \in \underline{\underline{\mathrm{Pic}}}_g(S^{(n+1)})$ by

$$\delta_{n-1}I = I_1 \otimes_{n+1} I_2^* \otimes_{n+1} \cdots \otimes_{n+1} I_{n+1}^{(*)}$$

Using the fact that $\Delta_n \circ \Delta_{n-1} : S^{(n)} \to S^{(n+2)}$ is the zero map, it follows that there exists a natural graded isomorphism

$$\lambda_I : \delta_n \delta_{n-1} I \longrightarrow S^{(n+2)}$$

Consider the category Ω with objects given by pairs (I, α), where $I \in \underline{\underline{\mathrm{Pic}}}_g(S^{(n)})$, $\alpha : \delta_{n-1}I \to S^{(n+1)}$ is a graded isomorphism, and such that $\delta_n \alpha = \lambda_I$.

A morphism from (I, α) to (J, β) is given by a graded $S^{(n-1)}$-module homomorphism $f : I \to J$ such that $\alpha = \beta \circ \delta_{n-1} f$. The set of isomorphism classes of Ω forms an abelian group under the operation induced by the tensor product. Divide out the subgroup generated by classes of the form $(\delta_{n-2}J, \lambda_J)$, and call the quotient group E_n. For a purely K-theoretic definition of E_n, we refer to [72].

V.5.2 Lemma Let S be a commutative faithfully flat graded R-algebra, and let E_n be defined as above. Then $\mathrm{Pic}_g(R) \cong E_1$, and there exists a monomorphism $i : \mathrm{Br}_g(S/R) \to E_2$, which is an isomorphism if S is faithfully projective as an R-module.

Proof: For I representing $[I] \in \mathrm{Pic}_g(R)$, we define $(S \otimes I, \alpha) \in E_1$, where α is the canonical isomorphism induced by the switch map $S \otimes I \otimes S \to S \otimes S \otimes I$. This defines on isomorphism $\mathrm{Pic}_g(R) \cong E_1$, as one easily verifies using the theory of faithfully flat descent.

Suppose $[A] \in \mathrm{Br}_g(S/R)$; then there exists a graded isomorphism $\sigma : S \otimes A \to \mathrm{END}_S(Q)$. $\Phi = \sigma_2^{-1}\sigma_3\sigma_1$ is induced by a graded isomorphism $f : Q_1 \otimes_2 I_1 \to Q_2$, for some $I \in \underline{\underline{\mathrm{Pic}}}_g(S^{(2)})$. As $\varphi_2^{-1}\varphi_3\varphi_1 = 1$ is induced by $f_2^{-1} f_3 f_1$, it follows by III.1.7 that there exists a graded isomorphism $\alpha : I_2^* \times_3 I_3 I_1 \to S^{(3)}$. It is therefore easily established that (I, α) determines an element of

E_2. We leave it to the reader to verify that this establishes a well-defined monomorphism $\mathrm{Br}_g(S/R) \to E_2$. Actually the methods involved here are just the same as those used in the proof of III.5.5. Now let S be faithfully projective as an R-module, and take (I, α) in E_2. Then we may define

$$f = \alpha \otimes 1 : I_1 \otimes_3 I_3 \longrightarrow I_2$$

and $f_4^{-1} f_2^{-1} f_3 f_1$ is an automorphism of $I_{11} \otimes_4 I_{31} \otimes_4 I_{33}$, which is the identity, as $(I, \alpha) \in E_2$. Let Q be the S-module I, where S acts on the first factor, and consider

$$g : Q_1 \otimes_2 I = 0_1 \otimes_3 I_3 \xrightarrow{f} I_2 \xrightarrow{\tau_1} Q_2$$

Then we may check (cf. for example, [36]), that g induces a descent datum γ: $\mathrm{END}(Q_1) \to \mathrm{END}(Q_2)$, defining the required graded Azumaya algebra. ■

We come back to the exact sequences V.1.4. Let S be a graded étale covering of R and consider the functors

$$i : \underline{S}(R_{\text{ét}_g}) \longrightarrow \underline{P}(R_{\text{ét}_g})$$

$$g = H^0(S/R, \circ) : \underline{P}(R_{\text{ét}_g}) \longrightarrow \underline{Ab}$$

Then i takes injective sheaves to g-acyclic presheaves; indeed, $R_g^n = H^n(S/R), \cdot)$, and $H^n(S/R, iF) = 1$ if F is an injective sheaf, as follows from V.2.10 or from [44, III.2.4]. Furthermore, $R^n(g \circ i) = H^n(R_{\text{ét}_g}, \cdot)$, and $R^n i(F) = H^n(\cdot_{\text{ét}_g}, F)$. Before writing down some exact sequences, let us state:

V.5.3 Lemma Suppose R is strongly graded on the graded étale site, and let C_q^0 be the image under i of the qth kernel of an injective resolution of U_0 in $\underline{S}(R_{\text{ét}_g})$. Then $H^{n-1}(S/R, C_1^0) \cong E_n$ for $n > 1$.

Proof: Let $1 \to U_0 \to F_0 \to F_1 \to \cdots$ be an injective resolution of U_0. Then we have an exact sequence of sheaves

$$0 \longrightarrow U_0 \longrightarrow E_0 \longrightarrow C_1^0 \longrightarrow 0$$

Let $c \in C_1^0(S^{(n)})$ represent an element of $H^{n-1}(S/R, C_1^0)$. There exists a graded étale covering X of $S^{(n)}$ such that $\Delta c = \theta x$ for some $x \in F_0(X)$ in the following diagram with exact rows:

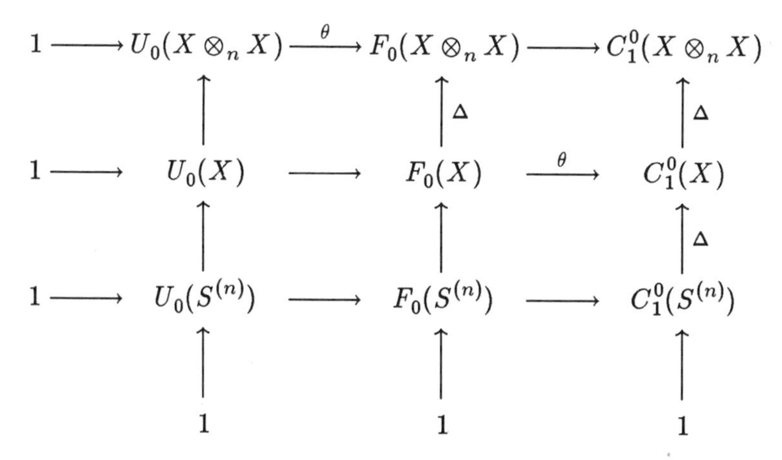

As $(\Delta \circ \theta)(x) = 1$, we may define $u = \theta^{-1}\Delta(x)$ and it is clear that $\Delta u = 1$, so multiplication by u yields a graded descent datum $I \in \underline{\underline{\mathrm{Pic}}}_g(S^{(n)})$. Denoting by δ the maps induced by $S^{(n)} \to S^{(n+1)}$, it follows from the fact that c is a cocycle that $(\theta \circ \delta)(x) = 1$, and therefore $\delta u \in \mathrm{Im}\,\Delta$, and we obtain a graded isomorphism $\alpha : \delta I \to S^{(n-1)}$. In similar way, we may deduce that $\delta\alpha = \lambda_I$, so (I, α) determines an element of E_n. The reader may check that this defines a monomorphism $i : H^{n-1}(S/R, C_1^0) \to E_n$.

To show that i is surjective, take $[(I, \alpha)] \in E_n$. Using the fact that R is strongly graded on the graded étale site, we obtain a graded étale covering X of $S^{(n)}$ splitting I in $\mathrm{Pic}_g(S^{(n)})$. From III.5.5, we know that I defines an element of $H^1(X/S^{(n)}, U_0)$, and then some diagram chasing in the above diagram yields an element of $C_1^0(S^{(n)})$ which is a cocycle. ∎

Now suppose that R is strongly graded on the graded étale site, and write down the exact sequences V.1.4 for $A = U_0$. Taking the preceding lemma into account, and the fact that $\mathrm{Pic}_g(R) =$

$H^1(R_{\text{ét}_g}, U_0)$ (V.3.11), we obtain

$$\begin{aligned} 1 &\longrightarrow H^1(S/R, U_0) \longrightarrow \text{Pic}_g(R) \longrightarrow H^0(S/R, \text{Pic}_g) \\ &\longrightarrow H^2(S/R, U_0) \longrightarrow E_2 \cong H^1(S/R, C_1^0) \\ &\longrightarrow H^1(S/R, \text{Pic}_g) \longrightarrow H^3(S/R, U_0) \longrightarrow \cdots \end{aligned}$$

for $q = 0$, and

$$\begin{aligned} 1 &\longrightarrow E_2 \cong H^1(S/R, C_1^0) \longrightarrow H^2(R_{\text{ét}_g}, U_0) \\ &\longrightarrow H^0(S/R, H^2(\cdot_{\text{ét}_g}, U_0)) \longrightarrow \cdots \end{aligned}$$

for $q = 1$. If S is faithfully projective, this is just the sequence III.5.5.

V.5.4 Corollary If R is strongly graded on the graded étale site, then we have a monomorphism $\text{Br}_g(R) \to H^2(R_{\text{ét}_g}, U_0)$.

Proof: From V.5.2, V.3.3 and the two exact sequences above. ■

Now let us turn to the general case. Choose injective resolutions for U and gr in the category $\underline{S}(R_{\text{ét}_g})$ as follows:

$$\begin{array}{ccccccccc} 1 & \longrightarrow & U & \longrightarrow & F_0 & \longrightarrow & F_1 & \longrightarrow & \cdots \\ & & \downarrow & & \downarrow & & \downarrow & & \\ 1 & \longrightarrow & \text{gr} & \longrightarrow & G_0 & \longrightarrow & G_1 & \longrightarrow & \cdots \end{array}$$

and let C_q and D_q be the images under i of their qth kernels; then we have a map

$$d : C_1 \longrightarrow D_1$$

Let E_2' be the subgroup of $H^1(S/R, C_1)$ consisting of those elements represented by a cocycle $c \in C_1(S \otimes S)$ satisfying $d(c) = 1$ in $D_1(S \otimes S)$. We have:

V.5.5 Proposition If S is a commutative graded faithfully flat R-algebra, then

$$E_2 \cong E_2' \subset \operatorname{Ker}(H^1(S/R, C_1) \longrightarrow H^1(S/R, D_1))$$

Proof: Let $c \in Z^1(S/R, C_1)$, and suppose $d(c) = 1$. Consider the diagrams (5.5.1) and (5.5.2):

$$\begin{array}{ccccccc}
1 \longrightarrow & U(X \otimes_2 X) & \longrightarrow & F_0(X \otimes_2 X) & \longrightarrow & C_1(X \otimes_2 X) & \\
& \uparrow & & \uparrow & & \uparrow & \\
1 \longrightarrow & U(X) & \longrightarrow & F_0(X) & \longrightarrow & C_1(X) & \\
& \uparrow & & \uparrow & & \uparrow & (5.5.1) \\
1 \longrightarrow & U(S^{(2)}) & \longrightarrow & F_0(S^{(2)}) & \longrightarrow & C_1(S^{(2)}) & \\
& \uparrow & & \uparrow & & \uparrow & \\
& 1 & & 1 & & 1 &
\end{array}$$

$$\begin{array}{ccccccc}
1 \longrightarrow & \operatorname{gr}(X \otimes_2 X) & \longrightarrow & G_0(X \otimes_2 X) & \longrightarrow & D_1(X \otimes_2 X) & \\
& \uparrow & & \uparrow & & \uparrow & \\
1 \longrightarrow & \operatorname{gr}(X) & \longrightarrow & G_0(X) & \longrightarrow & D_1(X) & \\
& \uparrow & & \uparrow & & \uparrow & (5.5.2) \\
1 \longrightarrow & \operatorname{gr}(S^{(2)}) & \longrightarrow & G_0(S^{(2)}) & \longrightarrow & D_1(S^{(2)}) & \\
& \uparrow & & \uparrow & & \uparrow & \\
& 1 & & 1 & & 1 &
\end{array}$$

All vertical maps are denoted by Δ, all horizontal ones by θ, and X is a graded étale covering of $S^{(2)}$ such that $\Delta(c) \in \operatorname{Im}\theta$, say $\Delta(c) = \theta(x)$. As in V.5.3, we find $u \in U(X \otimes_2 X)$ such that

$\theta(u) = \Delta(x)$. Now, as $d(c) = 1$, $d(x) \in \mathrm{Im}\,\theta$, say $d(x) = \theta(U)$, where $U \in \mathrm{gr}(X)$, and clearly $\Delta(U) = d(u)$. So multiplication by u yields graded isomorphisms

$$X \otimes_2 X \longrightarrow (X \otimes_2 X) \otimes \Delta(u)$$

and

$$X \otimes_2 U \longrightarrow U \otimes_2 X$$

which is a graded descent datum defining $I \in \underline{\underline{\mathrm{Pic}}}_g(S^{(2)})$. The unadorned symbol $\otimes$ means tensoring up on $X \otimes_2 X$ here.

Using the fact that C is a cocycle, we may show that $I_1 \otimes_3 I_3 \cong_g I_2$, and therefore we obtain a well-defined element of E_2. Routine verifications show that we have thus obtained a well-defined monomorphism $E_2' \to E_2$. To show that this monomorphism is surjective, take $[(I,\alpha)] \in E_2$, and let X be a graded étale covering of $S^{(2)}$ splitting I in $\mathrm{Pic}^g(S^{(2)})$. As in the proof of III.5.7, we obtain a cocycle $u \in U(X \otimes_2 X)$ such that $d(U) \in \mathrm{Im}\,\Delta$. Some diagram chasing yields a cocycle $c \in C_1(S^{(2)})$ such that $d(c) = 1$. ∎

As an application,. let us prove (cf. V.3.4):

V.5.6 Proposition Let R be a noetherian commutative graded ring; then the canonical morphism $\mathrm{Br}_g(R) \to H^2_{\mathrm{gr}}(R_{\text{ét}_g}, U)$ is injective.

Proof: Apply V.1.4 to the sequence of functors

$$\underline{S}(R_{'et_g}) \longrightarrow \underline{P}(R_{\text{ét}_g}) \longrightarrow \underline{Ab}$$

where we let $A = U$, resp. $A = \mathrm{gr}$. For $q = 0$, we obtain the diagram (5.6.1) with exact rows and columns. Recall that $H^1(R_{\text{ét}_g}, U) = \mathrm{Pic}^g(R)$, and $H^1(R_{\text{ét}_g}, \mathrm{gr}) = 1$. Some diagram chasing yields a sequence

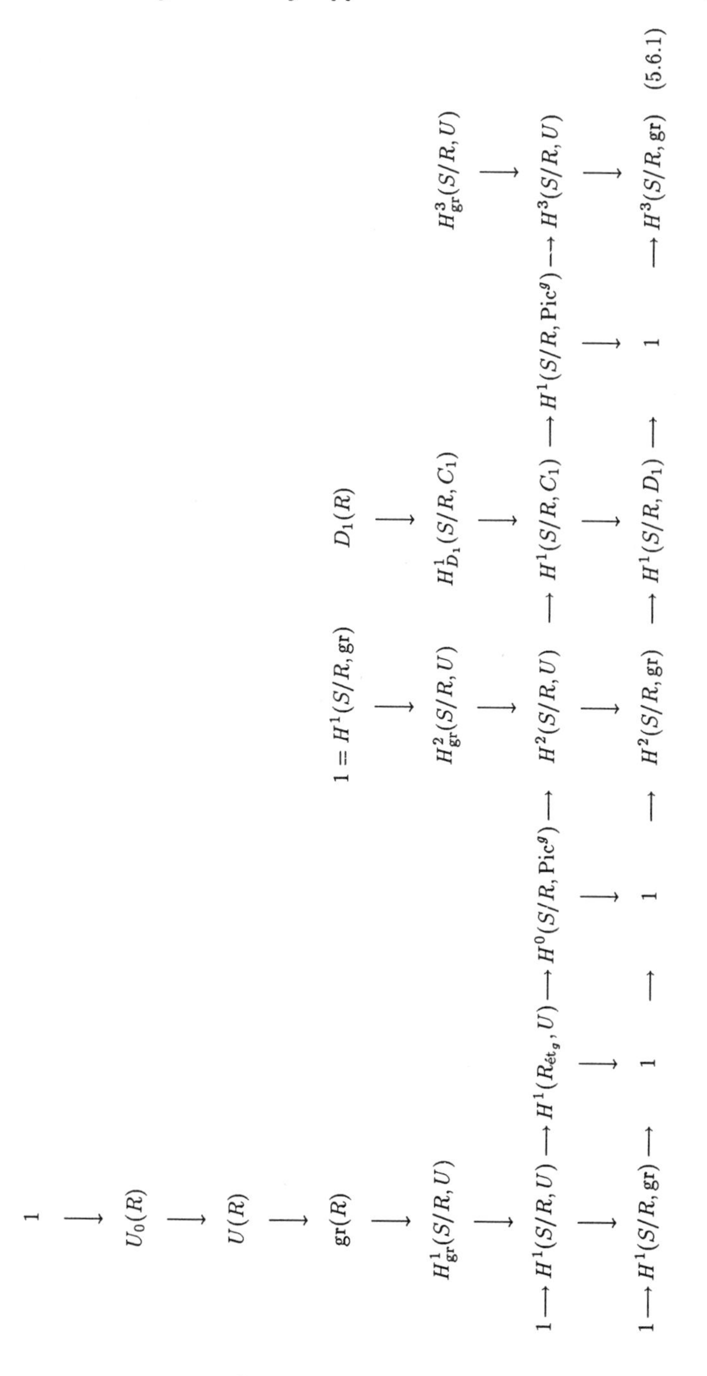

$$\begin{aligned}1 &\longrightarrow U_0(R) \longrightarrow U(R) \longrightarrow \mathrm{gr}(R) \longrightarrow H^1_{\mathrm{gr}}(S/R,U)\\ &\longrightarrow \mathrm{Pic}^g(R) \longrightarrow H^0(S/R,\mathrm{Pic}^g) \longrightarrow H^2_{\mathrm{gr}}(S/R,U)\\ &\longrightarrow \mathrm{Ker}(H^1(S/R,C_1) \longrightarrow H^1(S/R,D_1)\\ &\longrightarrow H^1(S/R,\mathrm{Pic}^g) \longrightarrow \mathrm{Ker}(H^3(S/R,U)\\ &\longrightarrow H^3(S/R,\mathrm{gr}))\end{aligned} \tag{5.6.2}$$

(5.6.2) is exact, except at $H^1(S/R,\mathrm{Pic}^g)$, and is nearly identical to III.5.8. For $q = 1$, we obtain

$$\begin{array}{ccccc} & & 1 & & \\ & & \downarrow & & \\ & & H^2_{\mathrm{gr}}(R_{\text{ét}_g},U) & & \\ & & \downarrow & & \\ 1 \longrightarrow H^1(S/R,C_1) & \longrightarrow & H^2(R_{\text{ét}_g},U) & \longrightarrow & H^0(S/R,H^2(\cdot,U)) \\ \downarrow & & \downarrow & & \downarrow \\ 1 \longrightarrow H^1(S/R,D_1) & \longrightarrow & H^2(R_{\text{ét}_g},\mathrm{gr}) & \longrightarrow & H^0(S/R,H^2(\cdot,\mathrm{gr})) \end{array} \tag{5.6.3}$$

Using V.5.2 and V.5.6, we obtain monomorphisms

$$\begin{aligned}\mathrm{Br}_g(S/R) \longrightarrow E_2 \longrightarrow \mathrm{Ker}(H^1(S/R,C_1)\\ \longrightarrow H^1(S/R,D_1)) \longrightarrow H^2_{\mathrm{gr}}(R_{\text{ét}_g},U)\end{aligned}$$

V.5.6 now follows from V.5.2. ∎

V.5.7 Corollary Let S be a commutative graded faithfully flat R-algebra. Then $H^1(S/R,\mathrm{gr}) = 1$, and consequently $\mathrm{Pic}(S/R) = \mathrm{Pic}^g(S/R) = H^1(S/R,U)$.

Proof: $H^1(S/R,\mathrm{gr}) = 1$ follows from (5.6.1). The rest follows from III.5.7 and [37, V.2.1]. ∎

V.5.8 Remark If R reduced, then gr is the constant sheaf $\mathbf{Z}$. In this case, D_1 may be computed as well. Indeed, $0 \to \mathbf{Z} \to \mathbf{Q} \to \mathbf{Q}/\mathbf{Z} \to 0$ is an injective resolution for $\mathbf{Z}$, so $D_1 = \mathbf{Q}/\mathbf{Z}$, and $D_n = 0$ for $n \geq 2$. From (5.6.1), it follows that $H^2(S/R,\mathbf{Z}) = H^1(S/R,\mathbf{Q}/\mathbf{Z})$. If S is a Galois extension of R with group G, this means that $H^2(G,\mathbf{Z}) = H^1(G,\mathbf{Q}/\mathbf{Z}) = \mathrm{Hom}(G,\mathbf{Q}/\mathbf{Z}) \cong G$.

As an example, let us compute the graded Brauer group of a Laurent polynomial ring.

V.5.9 Lemma Let S be an étale covering of R, where R is reduced, and let $M = R[T,T^{-1}]$, $S = N[T,T^{-1}]$, where $\deg T = n$. Then we have that

$$H^2_{\mathrm{gr}}(N/M,U) \cong H^2(S/R,U) \times H^1(S/R,\mathbf{Z}/n\mathbf{Z})$$

Proof: Let $x \in H^2_{\mathrm{gr}}(N/M,U)$ be represented by $(v,z) \in U(N^{(3)}) \times \mathbf{Z}(N^{(2)})$. We have that $v = T^{n_1}v_1e_1 + \cdots + T^{n_m}v_me_m$, where the e_i are orthogonal idempotents in $S^{(2)}$, $v_i \in S^{(2)}e_i$. Clearly $v_0 = v_1 + \cdots + v_m$ is a cocycle in $U(S^{(3)})$, and the map $\mathrm{Spec}(S^{(2)}) \to n\mathbf{Z} : \mathrm{Spec}(S^{(2)}e_i) \to nn_i$ determines the cocycle $d_2(v)$ (in the notation of III.5.4), and actually $d_2v = D_1z$. As d_2v takes values in $n\mathbf{Z}$, the image of d_2v in $\mathbf{Z}/n\mathbf{Z}$ is trivial; hence $\bar{z}$ represents a cocycle of $H^1(S/R,\mathbf{Z}/n\mathbf{Z})$. We set $\beta(x) = (v_0,\bar{z})$. It is easily seen that β is a homomorphism. Let us define an inverse for β. Take $(v_0,\bar{z})$ in $H^2(S/R,U) \times H^1(S/R,\mathbf{Z}/n\mathbf{Z})$, and take a representative z of $\bar{z}$ in $\mathbf{Z}(S^{(2)})$. Choose a component $S^{(2)}e$ of $S^{(2)}$ on which D_1z takes the constant value n_i. We define v by $ve = T^{n_i}v_0e_i$. It is clear that $\beta(v,z) = (v_0,\bar{z})$. So if we set $\gamma(v_0,\bar{z}) = (v,z)$, then γ is an inverse for β. ■

V.5.10 Theorem $\mathrm{Br}_g(R[T,T^{-1},\deg T = n]) = \mathrm{Br}(R) \times H^1(R_{\text{ét}},\mathbf{Z}/n\mathbf{Z})$.

Proof: For every graded M-Azumaya algebra A, there exists an étale covering $N = S[T,T^{-1}]$ splitting A. This holds for a graded field (IV.1.9), a gr-local ring (IV.2.4.2), and hence in general, by a local-global argument similar to V.3.3. We may now construct a monomorphism $\mathrm{Br}_g(M) \to \varinjlim H^2_{\mathrm{gr}}(N/M,U)$, where the limit is taken over étale coverings of the form $N = S[T,T^{-1}]$. Different approaches are possible; for instance, follow Knus and Ojanguren ([37, V.2]) and use Artin's lemma. The construction in this particular situation is carried out completely in [10]. Using V.5.9, we now obtain a monomorphism $\mathrm{Br}_g(M) \to H^2(R_{\text{ét}_g},U) \times H^1(R_{\text{ét}_g},\mathbf{Z}/n\mathbf{Z})$. Since $\mathrm{Br}(R) = H^2(R_{\text{ét}},U)_{\text{tors}}$, it is now sufficient to show that $f : {}_n\mathrm{Br}_g(M) \to H^1(R_{\text{ét}},\mathbf{Z}/n\mathbf{Z})$ is onto. In VI.1.9, we shall show that every element of $H^1(R_{\text{ét}},\mathbf{Z}/n\mathbf{Z})$ may be represented by an element of $H^1(S/R,\mathbf{Z}/n\mathbf{Z})$ for some Galois extension S of R. But then, using V.5.9 and III.5.8, we have a map $g : H^1(S/R,\mathbf{Z}/n\mathbf{Z}) \to H^2_{\mathrm{gr}}(S/R,U) \to \mathrm{Br}_g(S/R)$. Clearly $f(g(z)) = z$. ■

Remark It follows from the theorem above and from VI.1.9 that

$$\mathrm{Br}_g(R[T,T^{-1},\deg T = n]) = \mathrm{Br}(R) \times \mathrm{Gal}(R,\mathbf{Z}/n\mathbf{Z}).$$

V.5.11 Theorem Let $M = R[T,T^{-1},\deg T = n]$, and assume that $n^{-1} \in R$. Then $\mathrm{Ker}(\mathrm{Br}_g(R) \to \mathrm{Br}(R)) = C/nC$, where $C = \mathrm{Pic}(R[T,T^{-1}])/\mathrm{Pic}(R[T])$.

Proof: In [24], T. Ford showed that ${}_n\mathrm{Br}(M) = {}_n\mathrm{Br}(R) \times H^1(R_{\text{ét}},\mathbf{Z}/n\mathbf{Z})/(C/nC)$. A comparison between this result and V.6.2 yields the theorem. ■

Chapter VI

Applications

VI.1 THE BRAUER–LONG GROUP

In this section we use the notation introduced in I.1.2. In [96], F. Long introduced a Brauer group of a commutative ring with respect to a finite abelian group G. This group generalized, in a sense, the Brauer–Wall group [101, 107], the graded Brauer group of Childs, Garfinkel, and Orzech [86, 87], and the equivariant Brauer group of Fröhlich and Wall [91]. We shall apply the techniques of graded cohomology, as developed in the $\mathbb{Z}$-graded case in Chapters III, IV, and V to this situation; as the reader may expect, the cohomology groups of the group G will show up in the picture. We shall obtain a cohomological description of Long's Brauer group, or at least of some of its subgroups; this will yield some cohomological proofs of classical results concerning the Brauer–Long group (cf. [77–81, 89, 90, 96, 98, 99]).

VI.1.1 Dimodule Theory Let R be a commutative ring and G a finite abelian group. We suppose that R is trivially graded, and that G acts trivially on R. A *G-module* M is an R-module with an action of G on M such that ${}^{\sigma\tau}m = {}^{\sigma}({}^{\tau}m), {}^{1}m = m$, for all $m \in M, \sigma, \tau \in G$. A *$G$-dimodule* M is a G-graded G-module M such that ${}^{\sigma}(M_\tau) \subset M_\tau$ for all $\sigma, \tau \in G$. A *G-module algebra* A is an R-algebra which is a G-dimodule such that ${}^{\sigma}(ab) = {}^{\sigma}a{}^{\sigma}b$, for all $\sigma \in G, a, b \in A$. A *G-dimodule algebra* is a G-graded algebra, which is a G-dimodule and a G-module algebra.

For two G-modules M, N, $\mathrm{Hom}_R(M,N)$ may be given the structure of a G-module by defining $({}^{\sigma}f)(m) = {}^{\sigma}(f({}^{\sigma^{-1}}m))$, for all $\sigma \in G$, $m \in M$, $f \in \mathrm{Hom}_R(M,N)$. If M, N are dimodules, then $\mathrm{Hom}_R(M,N)$ furnished with its dimodule structure will be denoted $\mathcal{HOM}_R(M,N)$. If we consider only the module structure, then we write $\mathcal{H}om_R(M,N)$.

The tensor product $M \otimes N$ is a G-module if we define ${}^{\sigma}(m \otimes n) = {}^{\sigma}m \otimes {}^{\sigma}n$. Consider two G-dimodule algebras A and B; then the tensor product is again a dimodule algebra. On $A \otimes B$, we define a new multiplication as follows: for $b_1 \in B_\sigma, a_1, a_2 \in A, b_2 \in B$, we let $(a_1 \otimes b_1)(a_2 \otimes b_2) = a_1{}^{\sigma}a_2 \otimes b_1 b_2$. This new dimodule algebra is denoted $A \# B$; we call it the *smash product* of A and B. Note that the multiplication rule above also holds for the composition of homomorphisms of dimodules, that is, $(f_1 \# f_2) \circ (g_1 \# g_2) = (f_1 \circ {}^{\sigma}g_1) \# (f_2 \circ g_2)$, where $f_i : N_i \to P_i, g_i : M_i \to N_i$ and $\deg f_2 = \sigma$. We write $f \# g$ instead of $f \otimes g$. If a and b are invertible elements or morphisms, and $\deg b = \sigma$, then $(a \# b)^{-1} = ({}^{\sigma}a)^{-1} \# b^{-1}$. Let $G(R) = \{\sigma : \mathrm{Spec}(R) \to G \mid f \text{ is continuous}\}$, where $\mathrm{Spec}(R)$ is furnished with the Zariski topology, and G with the discrete topology; clearly $\sigma \in G(R)$ is constant on a connected component of R. Also $\sigma \in G(R)$ acts on R as follows: take a component R_i of R on which σ takes a constant value σ_i, and let ${}^{\sigma}x = {}^{\sigma}i_x$, for all $x \in R_i$. An element or morphism is said to be *locally homogeneous* of degree $\sigma \in G(R)$ if for all $p \in \operatorname{Spec}(R)$, f_p is homogeneous of degree $\sigma(p)$. It follows immediately that the multiplication rules given above for homogeneous elements and morphisms also apply to locally homogeneous elements and morphisms. Given a dimodule algebra A, we define the G-opposite algebra $\bar{A}$ of A as follows: as an R-module, $\bar{A} = A$, but the multiplication is given by $a \cdot b = ({}^{\sigma}b)a$, where a is (locally) homogeneous of degree σ.

Denote $G^*(R) = G(R)^* = \operatorname{Hom}(G(R), U(R))$. Given $(\sigma, \sigma^*) \in G(R) \times G^*(R)$, we define a G-dimodule as follows. As an R-module, $R(\sigma, \sigma^*)$ is equal to R. For the gradation, consider a component R_i of R where σ reaches a constant value σ_i. On R_i, the gradation is given by $(R_i)_{\sigma_i} = R_i, (R_i)_\tau = 0$ for $\tau \neq \sigma_i$. The action is given by ${}^{\tau}x = \sigma^*(\tau)x$ for $x \in R$, $\tau \in G(R)$. Consider the identity map $i : R \to R(\sigma, \sigma^*)$. Clearly i is not a dimodule isomorphism; however, it is locally homogeneous of degree σ, and the action of G on i is given by $({}^{\tau}i)(x) = {}^{\tau}(i({}^{\tau^{-1}}x) = {}^{\tau}(i(x)) = \sigma^*(\tau)i(x)$. Therefore ${}^{\tau}i = \sigma^*(\tau)i$, and, similarly, ${}^{\tau}i^{-1} = \sigma^*(\tau^{-1})i^{-1}$.

VI.1.2 The Brauer–Long Group of a Commutative Ring

Let R be a commutative ring and G a finite abelian group. An R-G-dimodule algebra A is called an *R-G-Azumaya algebra* if the following properties hold:

(1.2.1) A is faithfully projective as an R-module.

(1.2.2) The maps $F : A \# \bar{A} \to \operatorname{END}_R(A), G : \bar{A} \# A \to (\operatorname{END}_R(A))^{\mathrm{opp}}$, given by $F_{a\#\bar{b}}(c) = a({}^{\sigma}c)b$ and $G_{\bar{a}\#b}(c) = ({}^{\tau}a)cb$ ($\deg b = \sigma, \deg c = \tau$), are isomorphisms of dimodule algebras.

It may be shown (cf. [96, 97]) that the smash product of two G-Azumaya algebras is a G-Azumaya algebra, and the endomorphism ring $\operatorname{END}_R(M)$ of a faithfully projective G-dimodule is a G-Azumaya algebra. Also, if A is a G-Azumaya algebra, then so is $\bar{A}$. Two G-Azumaya algebras are called Brauer equivalent (denoted $A \sim B$) if there exist faithfully projective G-dimodules M, N such that $A \# \mathrm{END}_R(M) \cong_d B \# \mathrm{END}_R(N)$. (We use the notation $\cong_d$ to indicate that an isomorphism is a dimodule isomorphism.) The relation $\sim$ is an equivalence relation on the set of dimodule isomorphism classes of G-Azumaya algebras, and the quotient set forms a group under the multiplication induced by the smashed product. The inverse class of $[A]$ is given by $[\bar{A}]$. Following [90], we call this group the *Brauer–Long group of R with respect to* G, and we denote it by $\mathrm{BD}(R, G)$.

VI.1.3 Remarks

(1.3.1) Observe that a G-Azumaya algebra is not necessarily an Azumaya algebra. In general, the subset of $\mathrm{BD}(R, G)$ consisting of classes represented by an Azumaya algebra is not even a subgroup. Also note that the Brauer–Long group of an algebraically closed field is not necessary trivial.
(1.3.2) In this section, the ground ring R is always trivially graded, and has trivial G-action.

VI.1.4 Subgroups of the Brauer–Long group

(1.4.1) Let S be a commutative R-algebra. Then the map $[A] \to [S \otimes A]$ defines a morphism $\mathrm{BD}(R) \to \mathrm{BD}(S)$; its kernel is denoted $\mathrm{BD}(S/R)$.
(1.4.2) Given an R-Azumaya algebra A, we can define a G-Azumaya algebra by giving A trivial action and grading. This defines a monomorphism $\mathrm{Br}(R) \to \mathrm{BD}(R, G)$. We look at $\mathrm{Br}(R)$ as a normal subgroup of $\mathrm{BD}(R, G)$.
(1.4.3) Consider the subgroup of $\mathrm{BD}(R, G)$ consisting of classes represented by a G-Azumaya algebra with trivial action. It is easily seen that on this subset, the smash product and the tensor product are equal, and that each element is represented by an R-Azumaya algebra. Therefore, it is easily seen that this set forms a group, which we will denote by $\mathrm{BC}(R, G)$. This group is the

equivalent of the $\mathbf{Z}$-graded Brauer group considered in the previous chapters.

(1.4.4) Similarly, $\mathrm{BM}(R,G)$ is the subgroup consisting of classes represented by a G-Azumaya algebra with trivial grading.

(1.4.5) Fix a bimultiplicative map $\varphi : G(R) \times G(R) \to U(R)$, and consider G-Azumaya algebras A such that the action on A is given by the formula

$$^{\gamma}a = \varphi(\gamma, \deg a)a$$

for homogeneous a. The subgroup of $\mathrm{BD}(R,G)$ consisting of classes represented by such an algebra is denoted by $B_{\varphi}(R,G)$, and we call it the *graded Brauer group of Childs, Garfinkel, and Orzech* (cf. [86, 87]). In the case where $G = C_2 = \{-1,1\}$, $\varphi(-1,-1) = -1$, and $\varphi(e_i,e_j) = 1$ otherwise, this group coincides with the Brauer–Wall group (cf. [109] in the case where R is a field, [101] in general); that is, $B_{\varphi}(R,\mathbf{Z}/2\mathbf{Z}) = \mathrm{BW}(R)$.

(1.4.6) $\mathrm{BD}^s(R,G)$ will be the subgroups of all classes in $\mathrm{BD}(R,G)$ split by some faithfully flat R-algebra S; that is, $\mathrm{BD}^s(R,G) = \cup \mathrm{BD}(S/R,G)$, where the union is over all faithfully flat R-algebras. For a subgroup X of $\mathrm{BD}(R,G)$ we denote $X^s = X \cap \mathrm{BD}^s(R,G)$.

(1.4.7) Further, we shall see that not every class represented by an R-Azumaya algebra can be split by a faithfully flat extension. The set of classes represented by a G-Azumaya algebra which is Azumaya will be denoted $\mathrm{BAz}(R,G)$. In general, $\mathrm{BAz}(R,G)$ is not a group (cf. [98]).

(1.4.8) We have:

$$\mathrm{Br}(R) \subset \begin{cases} \mathrm{BM}^s(R,G) \\ \mathrm{BC}^s(R,G) \end{cases} \subset \mathrm{BD}^s(R,G)$$

$$\subset \mathrm{BAz}(R,G) \subset \mathrm{BD}(R,G)$$

and

$$\mathrm{Br}(R) \subset B_{\varphi}(R,G) \subset \mathrm{BD}(R,G)$$

VI.1.5 Notations We denote by $\underline{\underline{FPD}}(R,G)$ the category of faithfully projective G-modules and dimodule homomorphisms; $\underline{\underline{PD}}(R,G)$ will be the category of invertible G-R-dimodules and dimodule homomorphisms; $\mathrm{PD}(R,G) = K_0\underline{\underline{PD}}(R,G)$ is the group of isomorphism classes of invertible dimodules. The tensor product is a product on both categories. We use similar notation for the subcategories $\underline{\underline{FPC}}(R,G)$, $\underline{\underline{PC}}(R,G)$, $\underline{\underline{FPM}}(R,G)$, and $\underline{\underline{PM}}(R,G)$ consisting of dimodules with, respectively, trivial action and grading. In the next lemma, we calculate $\mathrm{PD}(R,G)$.

VI.1.6 Lemma Let R be a commutative ring, and let R be a finite abelian group.
(1.6.1) $\mathrm{PD}(R,G) \cong \mathrm{Pic}(R) \times G(R) \times G^*(R)$.
(1.6.2) $\mathrm{PC}(R,G) \cong \mathrm{Pic}(R) \times G(R)$.
(1.6.3) $\mathrm{PM}(R,G) \cong \mathrm{Pic}(R) \times G^*(R)$.

Proof: Using the notation introduced in VI.1.1, we obtain a monomorphism $i : G(R) \times G^*(R) \to \mathrm{PD}(R,G) : (\sigma,\sigma^*) \to R(\sigma,\sigma^*)$. Next, consider an invertible R-dimodule I, and $p \in \mathrm{Spec}(R)$. As an R_p-module, $I_p \cong R_p$, and therefore I_p is just R_p as an R-module, but with gradation given by $\deg(x) = \sigma_p$ for some $\sigma_p \in G$, and for all $x \in I_p$. The map $\mathrm{Spec}(R) \to G : p \to \sigma_p$ is a well-defined element of σ of $G(R)$. An element σ^* of $G^*(R)$ is now defined as follows: for $p \in \mathrm{Spec}(R)$, we let 1_p be the image of 1 under the isomorphism $R_p \cong I_p$. For $\tau \in G(R)$, we define $\sigma^*(\tau)$ by $\sigma^*(\tau)_p = {}^{\tau_p}(1_p)$. Because τ is locally constant, it is easily seen that $\sigma^*(\tau)_p$ globalizes to $\sigma^*(\tau) \in R$. The map $(e,e^*) : \mathrm{PD}(R,G) \to G(R) \times G^*(R) : I \to (\sigma,\sigma^*)$ is an epimorphism split by i, and $\mathrm{Ker}(e,e^*) = \mathrm{Pic}(R)$. This proves (1.6.1), and the two other statements follow immediately. For $I \in \underline{\underline{PD}}(R,G)$, we denote $I = \underline{I} \otimes R(\sigma,\sigma^*)$, where $\underline{I} \cong I$ as an R-module, but with trivial action and gradation. Also $(\sigma,\sigma^*) = (e,e^*)(I)$. ∎

VI.1.7 Lemma (Faithfully Flat Descent for Dimodules) Let S be a faithfully flat commutative R-algebra, and M a G-dimodule. Suppose that $u : M_1 \to M_2$ is an $S^{(2)}$-G-dimodule isomorphism, and a descent datum, that is, $u_2 = u_3 u_1$. Then there exists an R-G-dimodule N and a dimodule isomorphism $\eta : N \otimes S \to M$ such that the following diagram of $S^{(2)}$-dimodule isomorphisms commutes (as usual, τ is the switch map):

$$\begin{array}{ccc} N_{13} & \xrightarrow{\eta_1} & M_1 \\ \downarrow{\scriptstyle \tau_3} & & \downarrow{\scriptstyle u} \\ N_{23} & \xrightarrow{\eta_2} & M_2 \end{array}$$

The pair (N,η) is unique up to a dimodule isomorphism. If M is a dimodule algebra, then L has s unique R-dimodule algebra structure such that η is an S-dimodule algebra isomorphism. Similar statements are valid for G-modules and G-graded modules.

Proof: This follows immediately from III.1.3; it is easily seen that the descended module is a dimodule. ■

VI.1.8 Lemma (Morita) Let $P,Q \in \underline{\underline{FPD}}(R,G)$ and $\alpha : \mathrm{END}_R(P) \to \mathrm{END}_R(Q)$ an R-G-dimodule algebra isomorphism. Then there exists $I \in \underline{\underline{PD}}(R,G)$ and a dimodule isomorphism $f : P \otimes I \to Q$ such that f induces α, that is, $\alpha(h) = f(h \otimes 1)f^{-1}$. Then f and I are unique up to a dimodule isomorphism. Similar statements hold for G-modules and G-graded modules.

Proof: Observe that the Morita theorems [23, I.3.3] are easily generalizable to the corresponding categories of dimodules. Then the proof of the lemma is a straightforward generalization of [37, IV.3.1]. ■

As an application of the two previous lemmas, let us compute the group of Galois extensions of R with given abelian group G. Recall (cf. [23, II.2]) that a Galois extension S of R with group G is an R-algebra which is a G-module such that $S^G = R$, and the map $l : S \otimes S \to \nabla(S : G) = GS = \oplus_{\sigma \in G} S v_\sigma$, given by $l(a \otimes b) = \Sigma_{\sigma \in G} \sigma(a) b v_\sigma$, is an isomorphism. Note that l can be made into a G-module isomorphism as follows: let $\underline{S}$ be the R-algebra S, but with trivial action, and let G act on $\underline{S}G$ using the rule $\tau_v = v_{\sigma\tau^{-1}}$. Then $l : S \otimes \underline{S} \to \underline{S}G$ is a G-module algebra isomorphism. The set of isomorphism classes of Galois extensions with (abelian) group G will be denoted $T(R,G)$.

Following [92], we define a group structure on $T(R,G)$. We remark that RG itself is a Galois extension of R; also, if $S \in T(R,G), T \in T(R,H)$, then $S \times T \in T(R, G \times H)$, and for a normal subgroup H of G and $S \in T(R,G)$, we have that $S^H \in T(R,H)$ (cf. [23, III.1.1]). Consider a group homomorphism $\varphi : G \to H$; then we have a morphism

$$T(R,\varphi) : T(R,G) \longrightarrow T(R,H)$$

defined by

$$T(R,\varphi)(S) = (S \times RH)^K$$

where $K = \mathrm{Ker}(G \times H \to H : (\sigma,\rho) \to \varphi(\sigma)\rho))$. Now the map

$$T(R,G) \times T(R,G) \longrightarrow T(R, G \times G) \xrightarrow{T(R,m)} T(R,G)$$

where m is the multiplication map, defines a group structure on $T(R,G)$ (cf. [92]).

VI.1.9 Theorem Let R be a commutative ring, G a finite abelian group. Then $T(R,G) \cong H^1(R_{\text{ét}}, G) \cong H^1(R_{\text{fl}}, G)$. Consequently every element of $H^1(R_{\text{ét}}, G)$ may be represented by a cocycle in $H^1(S/R, G)$ for some faithfully projective étale extension S of R.

Proof: Take $S \in T(R,G)$, and define Φ by commutativity of the following diagram:

$$\begin{array}{ccc} S_{13} = \underline{S} \otimes S \otimes \underline{S} & \xrightarrow{l_1} & (G\underline{S})_1 = G(\underline{S}^{(2)}) \\ \downarrow \tau_3 & & \downarrow \Phi \\ S_{23} = S \otimes \underline{S} \otimes \underline{S} & \xrightarrow{l_2} & (G\underline{S})_2 = G(\underline{S}^{(2)}) \end{array}$$

It is easily seen that Φ is a G-module algebra isomorphism. Now, $\Phi(v_e)$ is an idempotent in $G(\underline{S}^{(2)})$; hence it is of the form

$$\Phi(v_e) = \Sigma_{\sigma \in G} a_\sigma v_\sigma$$

where

$$1 = \Sigma_{\sigma \in G} a_\sigma$$

in $S^{(2)}$, and the a_σ are idempotents in $S^{(2)}$. The map u : $\mathrm{Spec}(S^{(2)}) \to G$ given by $u(\mathrm{Spec}(S^{(2)}a_\sigma) = \sigma$ determines an element of $G(S^{(2)})$, which is a cocycle because Φ is a descent datum. We map S to the class $[u]$ in $H^1(R_{\text{ét}}, G) = \varinjlim H^1(S/R,G)$ (we use the fact that a Galois extension is an étale covering). This defines maps $\eta : T(R,G) \to H^1(R_{\text{ét}}, G) \to H^1(R_{\text{fl}}, G)$. Let us show that η is a homomorphism. First, observe that if H is a subgroup of G, then S^H maps to $\bar{v}$ if S maps to v; that is, the following diagram is commutative:

$$\begin{array}{ccc} T(R,G) & \longrightarrow & T(R,(G/H)R) \\ \downarrow & & \downarrow \\ H^1(R_{\text{ét}}, G) & \longrightarrow & H^1(R_{\text{ét}}, G/H) \end{array} \qquad \begin{array}{ccc} S & \longrightarrow & S^H \\ \downarrow & & \downarrow \\ v & \longrightarrow & \bar{v} \end{array}$$

Next, if $\varphi : G \to H$ is a group homomorphism, then the following diagram is commutative:

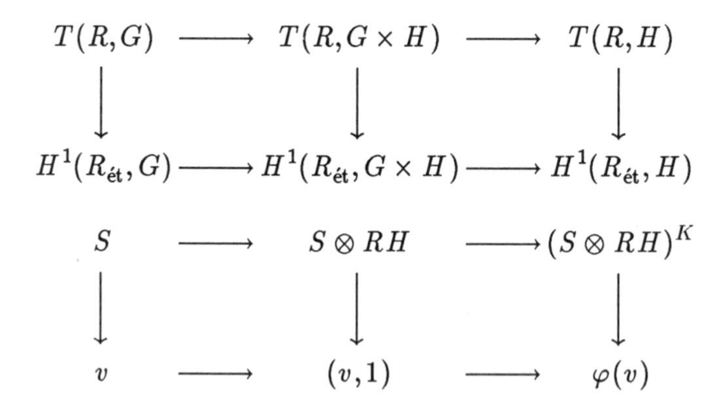

Indeed, $G \times H/K \cong H$, and the class $\overline{(v,1)} = \overline{(v,1)}\overline{(v^{-1},\varphi(v))} = \overline{(1,\varphi(v))}$, so $(v,1)$ maps to $\varphi(v)$. The fact that η is a homomorphism follows easily now, from the definition of the multiplication on $T(R,G)$. Also, the injectivity of η follows from the uniqueness in VI.1.7.

Let us prove that η is surjective. Represent an element of $H^1(R_{\text{fl}},G)$ by $u \in Z^1(T/R,G)$ for some flat covering T of R. As above, u corresponds to a map $\Phi : G(T^{(2)}) \to G(T^{(2)})$, which is a descent datum: GT descends to an R-algebra S with G-action. Let us prove that $S \otimes \underline{S} \cong G\underline{S}$. First, note that $GT \otimes \underline{S}$ descends to $S \otimes \underline{S}$. But $GT \otimes \underline{S} = G(T \otimes \underline{S}) = G(\underline{G(T)})$ also descends to $G\underline{S}$, and this proves the statement (using the uniqueness in VI.1.7). It follows that S is Galois, and clearly $\eta(S) = [u]$.

Finally, consider $[u] \in H^1(R_{\text{fl}},G)$: u represents $S \in T(R,G)$, and is represented by an element of $H^1(\underline{S}/R,G)$. Further, $\underline{S}$ is faithfully projective and étale. ■

We introduce the dual notion of a Galois extension. A Galois* extension of R with (abelian) group G is a G-graded commutative R-algebra S, which is faithfully flat as an R-algebra and which is such that the map $k : S \otimes \underline{S} \to \underline{S}G$, given by $k(a \otimes b) = \oplus_{\sigma \in G} a_\sigma b u_\sigma$, is a graded isomorphism. $G\underline{S}$ is the group ring with basis $\{u_\sigma : \sigma \in G\}$, and a_σ is the homogeneous part of degree σ of a. Also, $\underline{S}$ is just S, but with trivial gradation.

Clearly, RG is a Galois* extension of R itself, and the tensor product induces a map

$$T^*(R,G) \times T^*(R,H) \longrightarrow T^*(R,G \times H)$$

where $T^*(R,G)$ is the set of isomorphism classes of Galois* extensions. For a homomorphism $\varphi : H \to G$ of abelian groups, we can define a map $T^*(R,\varphi) : T^*(R,G) \to T^*(R,H)$ as follows: for $S \in T^*(R,G)$, let $S^{(H)}$ be the H-graded R-algebra given by $(S^{(H)})_\sigma = S_{\varphi(\sigma)}$. The map $G \to G \times G : \sigma \to (\sigma,\sigma)$ defines a multiplication map $T^*(R,G) \times T^*(R,G) \to T^*(R,G)$; that is, the multiplication ST of S and T is given by $ST = \oplus_{\sigma \in G} S_\sigma \otimes T_\sigma \subset S \otimes T$.

VI.1.10 Theorem Let R be a commutative ring, and let G be a finite abelian group. Then $T^*(R,G) \cong H^1(R_{\text{fl}}, G^*)$. Consequently every element of $H^1(R_{\text{fl}}, G^*)$ may be represented by a cocycle in $H^1(S/R, G^*)$ for some faithfully projective extension S of R.

Proof: Take $S \in T^*(R,G)$, and define Φ by commutativity of the following diagram of graded $\underline{S}^{(2)}$-algebra isomorphisms:

$$\begin{array}{ccc}
\underline{S} \otimes S \otimes \underline{S} = S_{13} & \xrightarrow{1 \otimes k} & \underline{S} \otimes \underline{S}G \cong (\underline{S} \otimes \underline{S})G \\
\downarrow & & \downarrow \Phi \\
S \otimes \underline{S} \otimes \underline{S} = S_{23} & \xrightarrow{k \otimes 1} & \underline{S}G \otimes \underline{S} \cong (\underline{S} \otimes \underline{S})G
\end{array}$$

Φ is gradation preserving, so $\Phi(u_\sigma) = a_\sigma u_\sigma$, for some $a_\sigma \in \underline{S}^{(2)}$. Because Φ is an algebra isomorphism, the map $\sigma \to a_\sigma$ is a homomorphism $G \to U(\underline{S}^{(2)})$; therefore, it determines an element of

$G^*(\underline{S}^{(2)})$, which is a cocycle, since Φ is a descent datum. To finish the proof, we continue as before. ■

VI.1.11 Remarks

(1.11.1) A Galois* extension is not necessarily separable, and this is the reason, in general, we have to work on the flat site instead of the étale site. An example of a nonseparable Galois* extension is given by $\mathbf{Z}[i]/\mathbf{Z}$.

(1.11.2) If R has a primitive nth root of unity in every component, and n is invertible in R ($n = \exp G$), then the sheaves G and G^* are isomorphic. Galois and Galois* extensions then coincide, and they obtain a dimodule structure. For instance, if G is a cyclic group with generator σ, then the G-action and G-gradation define each other by the following relation: ${}^{\sigma}a = \omega^i a$ if $\deg a = \sigma^i$ (ω being a primitive nth root of unity). In this situation we also have that a Galois extension is a G-Azumaya algebra (which is not Azumaya!). Indeed, $\underline{S} \otimes S = \underline{S}G$ is an $\underline{S}$-G-Azumaya algebra (check the definition), and then it follows that S is G-Azumaya by VI.1.7.

We now define a generalization of the cup product map. Let S be a faithfully flat R-algebra, and consider the map φ : $G^*(S^{(2)}) \times G(S^{(2)}) \to U(S^{(3)})$ given by $\varphi(\sigma^*, \sigma) = \sigma_1^*(\sigma_3)$. Recall that $\sigma_i = G(\varepsilon_i)(\sigma)$, $\sigma_i^* = G^*(\varepsilon_i)(\sigma^*)$.

VI.1.12 Lemma The map φ defined above induces well-defined maps (still denoted by φ):

$$\varphi : H^1(S/R, G^*) \times H^1(S/R, G) \longrightarrow H^2(S/R, U)$$

and

$$\varphi : H^1(R, G^*) \times H^1(R, G) \longrightarrow H^2(R, U),$$

where the cohomology may be taken on the flat or étale site.

Proof: First, we show that if σ^* and σ are cocycles, then $\varphi(\sigma^*,\sigma)$ is a cocycle in $U(S^{(3)})$. Indeed,

$$\begin{aligned}
\Delta_2\varphi(\sigma^*,\sigma) &= \sigma_1^*(\sigma_3)_1\sigma_1^*(\sigma_3)_2^{-1}\sigma_1^*(\sigma_3)_3\sigma_1^*(\sigma_3)_4^{-1} \\
&= \sigma_{11}^*(\sigma_{31})\sigma_{12}^*(\sigma_{32}^{-1})\sigma_{13}^*(\sigma_{33})\sigma_{14}^*(\sigma_{34}^{-1}) \\
&= \sigma_{11}^*(\sigma_{31}\sigma_{32}^{-1})(\sigma_{13}^*(\sigma_{14}^*)^{-1})(\sigma_{34}) \\
&\qquad\qquad (\text{because } \sigma_{11}^* = \sigma_{12}^*, \sigma_{33} = \sigma_{34}) \\
&= \sigma_{11}^*((\sigma_1\sigma_2^{-1})_4)(\sigma_2^*\sigma_3^{*-1})_1(\sigma_{34}) \\
&= \sigma_{11}^*(\sigma_{34}^{-1})\sigma_{11}^*(\sigma_{34}) \qquad (\sigma \text{ and } \sigma^* \text{ are cocycles}) \\
&= 1.
\end{aligned}$$

Also, if σ^* or σ is a coboundary, then so is $\varphi(\sigma^*,\sigma)$. For example, let $\sigma = \tau_1^{-1}\tau_2$ for some $\tau \in G(S)$; then

$$\begin{aligned}
\Delta_1(\sigma^*)^{-1}(\tau_2) &= ((\sigma^*)^{-1}(\tau_2))_1((\sigma^*)^{-1}(\tau_2))_2^{-1}((\sigma^*)^{-1}(\tau_2))_3 \\
&= (\sigma_1^*)^{-1}(\tau_{21})\sigma_2^*(\tau_{22})(\sigma_3^*)^{-1}(\tau_{23}) \\
&= (\sigma_1^*)^{-1}(\tau_{13})(\sigma_2^*(\sigma_3^*)^{-1})(\tau_{23}) \qquad (\tau_{22} = \tau_{23}) \\
&= (\sigma_1^*)(\tau_{13}^{-1})(\sigma_1^*(\tau_{23})) \\
&= \sigma_1^*((\tau_1^{-1}\tau_2)_3) = \sigma_1^*(\sigma_3) = \varphi(\sigma^*,\sigma).
\end{aligned}$$

The first part of the lemma is proved. For the second part, observe that we have some maps $H^1(R,G^*) \times H^1(R,G) \cong \varinjlim H^1(S/R,G^*) \times H^1(S/R,G) \xrightarrow{\varphi} \varinjlim H^2(S/R,U) \to H^2(R,U)$, where the limits are taken over all étale or flat coverings. ■

VI.1.13 Definition As in [75] and Section V.5, we define abelian groups E_1, E_2 for the Brauer–Long group. Let S be a faithfully flat R-algebra, take $I \in \underline{\underline{PD}}(S^{(n)},G)$, and define $\delta_{n-1}I \in \underline{\underline{PD}}(S^{(n+1)},G)$ as follows:

$$\delta_{n-1}I = I_1 \#_{n+1} I_2^* \#_{n+1} \cdots \#_{n+1} I_n^{(*)}$$

Because $\Delta_n \circ \Delta_{n-1} : S^{(n)} \to S^{(n+2)}$ is the zero map, it follows that we have a natural isomorphism of dimodules

$$\lambda_I : \delta_n\delta_{n-1}I \longrightarrow S^{(n+2)}$$

Now consider the category Ω with objects given by pairs (I,α), where $I \in \underline{\underline{PD}}(S^{(n)},G)$ and $\alpha : \delta_{n-1}I \to S^{(n+1)}$ is a dimodule isomorphism such that $\delta_n\alpha = \lambda_I$. A morphism between (I,α) and (J,β) in Ω is given by an $S^{(n+1)}$-dimodule homomorphism $f : I \to J$ such that $\alpha = \beta \circ \delta_{n-1}f$. Take the set Z_n of isomorphism classes of Ω. For $n = 1$, Z_n is made into a group by just taking the tensor product. For $n = 2$, we consider the following, more complicated, multiplication rule:

$$(I,\alpha)(J,\beta) = (I \#_2 J, (\alpha \# \beta)\varphi(e^*(I), e(J))$$

where e^* and e are defined as in VI.1.6. It is at this point that our approach is different from the classical one; in fact, it will turn out that the factor $\varphi(e^*(I), e(J))$ is due to the twisted multiplication in the smash product. We define $E_1 = Z_1$ and $E_2 = Z_2/B_2$, where B_2 is the subgroup of Z_2 consisting of elements of the form $(\delta_0 J, \lambda_j)$. Using VI.1.12, we can show that B_2 is a group.

VI.1.14 Proposition Suppose that S is a faithfully flat R-algebra, and let E_1, E_2 be defined as above. Then $\mathrm{PD}(R,G) \cong E_1$, and there exists a monomorphism $i : \mathrm{BD}(S/R,G) \to E_2$, which is an isomorphism if S is faithfully projective as an R-module.

Proof: The proof of the first part is a straightforward application of VI.3.7 and VI.3.8. For the second part, take $[A] \in \mathrm{BD}(S/R,G)$. Then we have a dimodule isomorphism $\rho : A \otimes S \to \mathcal{END}_S(Q)$, for some $Q \in \underline{\underline{FPD}}(S,G)$. As usual, define Φ by commutativity of the following diagram:

$$\begin{array}{ccc} A_{13} & \xrightarrow{\rho_1} & \mathcal{END}_2(Q_1) \\ \Big\downarrow{\scriptstyle \tau_3} & & \Big\downarrow{\scriptstyle \Phi} \\ A_{23} & \xrightarrow{\rho_3} & \mathcal{END}_2(Q_2) \end{array}$$

By IV.2.8, Φ is induced by a dimodule isomorphism $f : Q_1 \#_2 I \to Q_2$, for some $I \in \underline{\underline{PD}}(S^{(2)}, G)$. Now

$$f_2^{-1} f_3 f_1 : Q_{11} \#_3 I_2^* \#_3 I_3 \#_3 I_1 \longrightarrow Q_{11}$$

induces $\Phi_2^{-1}\Phi_3\Phi_1 = 1$, so from the uniqueness in VI.1.8, it follows that we have an $S^{(3)}$-dimodule isomorphism $u : S^{(3)} \to \delta_1 I$. Consider the map

$$Q_{11} \longrightarrow Q_{11} \#_3 S^{(3)} \xrightarrow{I\#u} Q_{11} \#_3 \delta_1 I \xrightarrow{f_2^{-1} f_3 f_1} Q_{11}$$

which is given by multiplication by a unit $m \in U(S^{(3)})$. It follows that the map $f_2^{-1} f_3 f_1 : Q_{11} \#_3 \delta_1 I \to Q_{11} = Q_{11} \#_3 S^{(3)}$ is of the form $1 \# \mu^{-1} = 1 \# \alpha$. Define $i([A]) = (I, \alpha)$. The proof of the fact that i is a well-defined monomorphism is long but rather straightforward. In fact, the most interesting point for us is to show that i is a homomorphism, because this has to do with the special product on E_2. The rest is similar to the corresponding proof for the usual Brauer group.

So let $[A']$ be another element of BD(S/R,G), and let ρ', Q', Φ', f', I' be defined as above. Clearly $\Phi \# \Phi'$ is induced by

$$f \# f' : Q_1 \#_2 I \#_2 Q_1' \#_2 I' \longrightarrow Q_2 \#_2 Q_2'$$

If we want to calculate $\Delta_1(f \# f')$, then we have the problem that we cannot interchange I and Q_1'. To overcome this, we observe that Φ and Φ' are also induced by

$$\underline{f} : Q_1 \#_2 \underline{I} \longrightarrow Q_2 \qquad \text{and} \qquad \underline{f}' : Q_1' \#_2 \underline{I}' \longrightarrow Q_2'$$

Here $\underline{I}$ and $\underline{I}'$ have no action or grading, but $\underline{f}$ and $\underline{f}'$ are not dimodule isomorphisms anymore. From VI.1.6, it follows that $\underline{f}$ and $\underline{f}'$ are locally homogeneous of degree σ and σ', and $G(S^{(2)})$ acts on $\underline{f}$ and $\underline{f}'$ as follows:

$${}^{\tau}\underline{f} = \sigma^*(\tau)\underline{f} \qquad \text{and} \qquad {}^{\tau}\underline{f}' = \sigma'^*(\tau)\underline{f}'$$

where we denoted

$$(e, e^*)(I) = (\sigma, \sigma^*) \qquad \text{and} \qquad (e, e^*)(I') = (\sigma', \sigma'^*)$$

Since $\underline{I}$ has neither action nor grading, it is no problem to interchange $\underline{I}$ and Q'_1. Therefore, $\Phi \# \Phi'$ is induced by $\underline{f} \# \underline{f}'$, and

$$\begin{aligned}\Delta_1(\underline{f} \# \underline{f}') &= (\underline{f}_2 \# \underline{f}'_2)^{-1}(\underline{f}_3 \# \underline{f}'_3)(\underline{f}_1 \# \underline{f}'_1)\\ &= \sigma_2^*(\sigma'_2)\sigma_1^*(\sigma'_3)(\underline{f}_2^{-1} \# \underline{f}_2'^{-1})(\underline{f}_3\underline{f}_1 \# \underline{f}'_3\underline{f}'_1)\\ &= \sigma_2^*(\sigma'_2)\sigma_1^*(\sigma'_3)(\sigma_3^*\sigma_1^*)(\sigma_2'^{-1})(\Delta_1\underline{f} \# \Delta_1\underline{f}')\\ &= \sigma_1^*(\sigma'_3)(\Delta_1\underline{f} \# \Delta_1\underline{f}') \qquad (\text{because } \sigma_2^* = \sigma_3^*\sigma_1^*)\end{aligned}$$

We have shown that

$$i(A \# A') = (I \#_2 I', \varphi(e^*(I), e(I'))(\alpha \# \alpha'))$$

so i is a homomorphism. Let us show that i is surjective if S/R is faithfully projective. Take $(I,\alpha) \in E_2$. α may be viewed as an $S^{(3)}$-dimodule isomorphism $\alpha : I_1 \#_3 I_3 \to I_2$. Let P be the S-dimodule which is equal to I as a $\mathbf{Z}$-dimodule, but where S acts on the first factor, that is, $s \cdot x = (s \otimes 1)x$ for $s \in S, x \in I$. Then P is a faithfully projective S-module, and

$P_1 = I_1$, with $S^{(2)}$-action on the first two factors.
$P_2 = I_3$, with $S^{(2)}$-action on the first and the third factor.
$P_1 \#_2 I = I_1 \#_3 I_3$, with $S^{(2)}$-action on the first two factors.

Therefore, we obtain an isomorphism of $S^{(2)}$-dimodules

$$f : P_1 \#_2 I = I_1 \#_3 I_3 \xrightarrow{\alpha} I_2 \xrightarrow{\tau_1} P_2$$

Note that α and τ_1 are not $S^{(2)}$-dimodule isomorphisms, but their composition is an $S^{(2)}$-dimodule isomorphism. Further, $f_2^{-1}f_3f_1$ is just multiplication by a unit, and therefore the induced map $\Psi : \mathcal{END}_2(P_1) \to \mathcal{END}_2(P_2)$ is a dimodule descent datum, descending to a dimodule-Azumaya algebra A. ■

Consider the category of sheaves $\underline{S}(R_{\mathrm{fl}})$ on the flat site of R. Let

$$1 \longrightarrow U \longrightarrow F_0 \longrightarrow F_1 \longrightarrow F_2 \longrightarrow \cdots$$

be an injective resolution of U in this category (cf. V.1 or [44]). Let $C_q = \mathrm{Ker}(F_q \to F_{q+1})$ (thus $C_0 = U$). For any faithfully flat

R-algebra S, we have some exact sequences (V.1.4 or [72, 6.13]):

$$(1.14.1)\quad 1 \longrightarrow H^1(S/R,U) \xrightarrow{\alpha_1} H^1(R,U) \xrightarrow{\beta_1} H^0(S/R,H^1(\cdot,U))$$
$$\xrightarrow{\gamma_1} H^2(S/R,U) \xrightarrow{\alpha_2} H^1(S/R,C_1) \xrightarrow{\beta_2} H^1(S/R,H^1(\cdot,U))$$
$$\xrightarrow{\gamma_2} H^3(S/R,U) \xrightarrow{\alpha_3} H^2(S/R,C_1) \longrightarrow \cdots$$

$$(1.14.2)\quad 1 \longrightarrow H^1(S/R,C_1) \longrightarrow H^2(R,U) \longrightarrow H^0(S/R,H(\cdot,U))$$
$$\longrightarrow H^2(S/R,C_1) \longrightarrow H^1(S/R,C_2) \longrightarrow \cdots$$

In the next proposition, the map α_2 will be important; we shall also consider the map $\alpha_2 \circ \varphi : H^1(S/R,G^*) \times H^1(S/R,G) \to H^2(R,U) \to H^1(S/R,C_1)$.

VI.1.15 Proposition Let S/R be faithfully flat; then

$$E_2 \cong H^1(S/R,G\times G^*) \times_{\alpha_2\circ\varphi} H^1(S/R,C_1)$$

where the multiplication on the product of cohomology groups is given by

$$(\sigma,\sigma^*,c)(\sigma',\sigma'^*,c') = (\sigma\sigma',\sigma^*\sigma'^*,cc'\alpha_2(\varphi(\sigma^*,\sigma')))$$

Proof: We construct an isomorphism $\delta : H^1(S/R,G\times G^*) \times_{\alpha_2\circ\varphi} H^1(S/R,C_1) \to E_2$. Let E_2^{class} be the classical analog of E_2, as introduced in [75]. Clearly, E_2^{class} is a normal subgroup of E_2, and it is well known that we have an isomorphism $\underline{\delta} : H^1(S/R,C_1) \to E_2^{\text{class}}$. Also, for u representing an element of $H^2(S/R,U)$, we have that $\underline{\delta}(\alpha_2(u)) = (S^{(2)},m(u))$, where $m(u)$ is just multiplication by u.

If (σ,σ^*) is a cocycle in $Z^1(S/R,G\times G^*)$, then the identity $i : \delta_1 S^{(2)}(\sigma,\sigma^*) \to S^{(3)}$ is a dimodule isomorphism. We define

$$\delta(\sigma,\sigma^*,c) = \underline{\delta}(c)(S^{(2)}(\sigma,\sigma^*),i)$$

Then

$$\delta((\sigma,\sigma^*,c),(\sigma',\sigma'^*,c')) = \delta((\sigma\sigma',\sigma^*\sigma'^*,cc'\alpha_2(\varphi(\sigma^*,\sigma'))$$
$$= \underline{\delta}(c)\underline{\delta}(c')(S^{(2)},m(\varphi(\sigma^*,\sigma'))(S^{(2)}(\sigma\sigma',\sigma^*\sigma'^*),i)$$
$$= \underline{\delta}(c)\underline{\delta}(c')(S^{(2)}(\sigma,\sigma^*),i)(S^{(2)}(\sigma',\sigma'^*),i))$$

$$= \delta(\sigma,\sigma^*,c)\delta(\sigma',\sigma'^*,c')$$

so δ is a homomorphism. The fact that δ is an isomorphism follows easily from the fact that $\underline{\delta}$ is an isomorphism. ■

VI.1.16 Corollary We have a monomorphism

$$\theta : \mathrm{BD}^s(R,G) \longrightarrow H^1(R_{\mathrm{fl}}, G \times G^*) \times_\varphi H^2(R,U)$$

Proof: For any faithfully flat R-algebra S, we have (using VI.1.14, VI.1.15 and (1.14.1)):

$$\begin{aligned} \mathrm{BD}(S/R) \subset E_2 &= H^1(S/R, G \times G^*) \times_{\alpha_2 \circ \varphi} H^1(S/R, C_1) \\ &\longrightarrow H^1(R_{\mathrm{fl}}, G \times G^*) \times_\varphi H^2(R,U). \end{aligned}$$

We obtain the result if we take the union over all faithfully flat S.

VI.1.17 Corollary (Chase–Rosenberg exact sequences) Let S be a faithfully projective R-algebra. We have the following exact sequences:

$$\begin{aligned} (1.17.1)\quad 1 &\longrightarrow (G \times G^*)(R) \times H^1(S/R,U) \longrightarrow \mathrm{PD}(R,G) \\ &\longrightarrow H^0(S/R,\mathrm{Pic}) \longrightarrow H^1(S/R, G \times G^*) \\ &\quad \times_\varphi H^2(S/R,U) \\ &\longrightarrow \mathrm{BD}(S/R,G) \longrightarrow H^1(S/R,\mathrm{Pic}) \end{aligned}$$

$$\begin{aligned} (1.17.2)\quad 1 &\longrightarrow H^1(S/R,U) \longrightarrow \mathrm{PD}(R,G) \\ &\longrightarrow H^0(S/R,\mathrm{PD}(\cdot,G)) \longrightarrow H^2(S/R,U) \\ &\longrightarrow \mathrm{BD}(S/R,G) \longrightarrow H^1(S/R,\mathrm{PD}(\cdot,G)) \\ &\longrightarrow H^3(S/R,U) \end{aligned}$$

Proof: (1.17.1) follows easily from (1.14.1) and the results above. (1.17.2) however, cannot be derived using Leray spectral sequences or exact couples. One may easily adapt Knus's proof ([36]) of the classical Chase–Rosenberg sequence. This is straightforward, so as in Chapter III, we leave the details to the reader. ■

VI.1.18 Remark Observe the analogy with the situation encountered in the **Z**-graded case (III.5.5, III.5.7). $H^1(S/R, G \times G^*) \times_\varphi H^2(S/R, U)$ is the analog of $H^2_{\mathrm{gr}}(S/R, U)$. Indeed, following III.3.4, consider the natural functor

$$F : \underline{\underline{PD}}(R, G) \longrightarrow \underline{\underline{\mathrm{Pic}}}(R)$$

We have a K-theoretic exact sequence

$$K_1 \underline{\underline{PD}}(R, G) \longrightarrow K_1 \underline{\underline{\mathrm{Pic}}}(R) \longrightarrow K_1 \Phi F$$
$$\longrightarrow K_0 \underline{\underline{PD}}(R, G) \longrightarrow K_0 \underline{\underline{\mathrm{Pic}}}(R)$$

that reduces to

$$1 \longrightarrow U(R) \longrightarrow U(R) \xrightarrow{d} (G \times G^*)(R) \longrightarrow \mathrm{PD}(R, G)$$
$$\longrightarrow \mathrm{Pic}(R) \longrightarrow 1$$

where $d = 1$. The analog of III.5.4 in this situation is given by

$$\begin{array}{ccccccc} 1 \longrightarrow & U(S) & \longrightarrow & U(S^{(2)}) & \longrightarrow \cdots \\ & \downarrow 1 & & \downarrow 1 & \\ 1 \longrightarrow & (G \times G^*)(S) & \longrightarrow & (G \times G^*)(S^{(2)}) & \longrightarrow \cdots \end{array}$$

Clearly, we will have that $H^{n+1}_{G \times G^*}(S/R, U) = H^n(S/R, G \times G^*) \times H^{n+1}(S/R, U)$.

VI.1.19 Theorem
(1.19.1) $\mathrm{BC}^s(R, G) \cong H^1(R, G) \times H^2(R, U)_{\mathrm{tors}}$.
(1.19.2) $\mathrm{BM}^s(R, G) \cong H^1(R_{\mathrm{fl}}, G^*) \times H^2(R, U)_{\mathrm{tors}}$.
(1.19.3) $\mathrm{BD}^s(R, G) \cong H^1(R_{\mathrm{fl}}, G \times G^*) \times_\varphi H^2(R, U)_{\mathrm{tors}}$.

Proof: Using Gabber's theorem (cf. [26, 31, 39]), we obtain a monomorphism $H^2(R_{\text{ét}}, U)_{\mathrm{tors}} \cong H^2(R_{\mathrm{fl}}, U)_{\mathrm{tors}} \cong \mathrm{Br}(R) \subset \mathrm{BD}^s(R, G)$. It remains to be shown that $\mathrm{BD}^s(R, G) \to H^1(R_{\mathrm{fl}}, G \times G^*)$ is surjective. From VI.1.9 and VI.1.10 it follows that an element of $H^1(R_{\mathrm{fl}}, G \times G^*)$ may be represented by an element of $H^1(S/R, G \times G^*)$ for some faithfully projective R-algebra S. Then apply VI.1.14. Finally, (1.19.1) and (1.19.2) follow immediately from (1.19.3). ∎

In VI.1.19, the cohomology groups may be taken on the flat or étale site, unless otherwise indicated. If $\exp(G)$ is invertible in R, then we can use the étale site everywhere. Our results may also be applied to the graded Brauer group of Childs, Garfinkel, and Orzech. Let $P_\varphi(R,G)$ be the subgroup of $\mathrm{PD}(R,G)$, consisting of classes represented by a dimodule for which the action and gradation are connected by the formula ${}^\gamma a = \varphi(\gamma, \deg a)a$.

The map $\varphi : G \times G \to U(R)$ may be extended easily to a morphism of sheaves $\varphi : G \times G \to U$ on R_{fl}. We define a morphism of sheaves

$$\nu : G \longrightarrow G^* : \sigma \longrightarrow (\gamma \longrightarrow \varphi(\gamma,\sigma) = \nu(\sigma)\gamma)$$

Using VI.1.6, we obtain immediately that

$$\begin{aligned} P_\varphi(R,G) &\cong \{(I,\sigma,\sigma^*) \in \mathrm{Pic}(R) \times G(R) \times G^*(R) : \sigma^* = \nu(\sigma)\} \\ &\cong \mathrm{Pic}(R) \times G(R) \end{aligned}$$

Also, from the proof of VI.1.19, it follows that

$$\begin{aligned} B^s_\varphi(R,G) &\cong \{(\sigma,\sigma^*,v) \in H^1(R,G) \times H^1(R_{\mathrm{fl}},G^*) \\ &\qquad \times_\varphi H^2(R,U)_{\mathrm{tors}} : \sigma^* = \nu(\sigma)\} \\ &\cong H^1(R,G) \times_\psi H^2(R,U)_{\mathrm{tors}} \end{aligned}$$

where ψ is given by

$$\psi(\sigma,\tau) = \varphi(\nu(\sigma),\tau) = \nu(\sigma_1)(\tau_3) = \varphi(\tau_3,\sigma_1)$$

We summarize our results:

VI.1.20 Proposition Let $\varphi : G(R) \times G(R) \to U(R)$ be a bilinear map, where G is an arbitrary abelian group. Then
(1.20.1) $P_\varphi(R,G) \cong \mathrm{Pic}(R) \times G(R)$.
(1.20.2) $B^s_\varphi(R,G) \cong H^1(R,G) \times_\varphi H^2(R,U)_{\mathrm{tors}}$.
where $\psi(\sigma,\tau) = \varphi(\tau_3,\sigma_1)$.

We leave it as an exercise to the reader to write down Chase–Rosenberg exact sequences for BC, BM, and B_φ.

We have already noted that not every element of BD(R,G) represented by a central algebra lies in BD$^s(R,G)$. We even have that the subset of BD(R,G) of classes represented by R-central algebras, BAz(R,G), is not a group:

VI.1.21 Example (Orzech [98]) Let $G = C_2 \times C_2 = \{1,\sigma,\tau,\sigma\tau\}$, and let R be a commutative ring such that $2^{-1} \in R$. On $M_2(R)$, we define a G-gradation as follows: $\deg x_\alpha = \alpha$, where

$$x_1 = \begin{bmatrix} 1 & 0 \\ 0 & 1 \end{bmatrix} \qquad x_\sigma = \begin{bmatrix} 0 & 1 \\ -1 & 0 \end{bmatrix}$$

$$x_\tau = \begin{bmatrix} 1 & 0 \\ 0 & -1 \end{bmatrix} \qquad x_{\sigma\tau} = \begin{bmatrix} 0 & 1 \\ 1 & 0 \end{bmatrix}$$

An action is defined by

$$\sigma \begin{bmatrix} a & b \\ c & d \end{bmatrix} = \begin{bmatrix} a & -b \\ -c & d \end{bmatrix} \qquad \tau \begin{bmatrix} a & b \\ c & d \end{bmatrix} = \begin{bmatrix} d & c \\ b & a \end{bmatrix}$$

This makes $M_2(R)$ into a dimodule Azumaya algebra A (cf. [98]). It is now an easy exercise to show that $\bar{A}$ is commutative. Hence $[\bar{A}] = [A]^{-1} \notin$ BAz(R,G).

To overcome this difficulty, we restrict our attention to the case where $G = C_n$, that is, G is a cyclic group. The point is that we now have that $H^2(G,U(R))$ contains only abelian elements; this fact will be used frequently in the sequel.

VI.1.22 Proposition BAz(R,C_n) is a subgroup of the group BD(R,C_n).

Proof: We follow the proof of Proposition 1.1 in [78]. First, suppose that R is local, and let $[A]$, $[B] \in$ BAz(R,G). Suppose that f is a normalized cocycle in $H^2(G,U(R))$, and let B_f be the graded algebra which is equal to B as a graded module, but with multiplication defined by

$$x \cdot y = f(\deg x, \deg y)xy$$

If f is an abelian cocycle, then B_f is central. Also, we have that B_f is separable: if $e = \Sigma_i x_i \otimes y_i \in (B \otimes B^{\text{opp}})_1$ is a separability idempotent for B, with x_i homogeneous of degree σ_i, then $e' = \Sigma_i f(\sigma_i, \sigma_i^{-1})x_i \otimes y_i$ is a separability idempotent for B_f.

R is local, so $\text{Pic}(R) = 1$ and the elements of G act as inner automorphisms on A. Let $x_\sigma \in A$ be such that ${}^\sigma a = x_\sigma a x_\sigma^{-1}$, for all $a \in A$; choose $x_1 = 1$. Then $x_\tau^{-1}x_\sigma^{-1}x_{\sigma\tau}$ induces the trivial action, and therefore, it is an element of $Z(A) = R$. Thus $f(\sigma,\tau) = x_{\sigma\tau}x_\tau^{-1}x_\sigma^{-1}$ defines a normalized cocycle in $H^2(G, U(R))$, which is abelian, because G is cyclic. We claim that $j : A \# B \to A \otimes B_f$, defined by $j(a \# b) = ax_\sigma \otimes b$, for b homogeneous of degree σ, is an R-algebra isomorphism. Indeed, for $a, c \in A, b \in B_\sigma, d \in B_\tau$, we have:

$$\begin{aligned} j((a \# b)(c \# d)) &= j(a^\sigma c \# bd) \\ &= a^\sigma c x_{\sigma\tau} \otimes bd \\ &= a x_\sigma c x_\sigma^{-1} x_{\sigma\tau} \otimes bd \\ &= a x_\sigma c x_\tau x_\tau^{-1} x_\sigma^{-1} x_{\sigma\tau} \otimes bd \\ &= f(\sigma,\tau) a x_\sigma c x_\tau \otimes bd \\ &= (a x_\sigma \otimes b)(c x_\tau \otimes d) \\ &= j(a \# b) j(c \# d) \end{aligned}$$

It follows that $A \# B \cong A \otimes B_f$ is central. If R is not local, then $A \# B$ is central because every one of its localizations is central. ■

The argument given above also proves that $\text{BAz}(R, G)$ is a group under the condition that $\text{Pic}_m(R) = 1 (m = \exp G)$ and that $H^2(G, U(R))$ consists of abelian cocycles. We are now working towards a complete description of $\mathrm{BAz}(R, C_n)$. But first, we need a lemma.

VI.1.23 Lemma (Graded version of the Rosenberg–Zelinsky sequence) Let A be an R-Azumaya algebra which is G-graded. Then we have an exact sequence $1 \to \mathrm{Inn}_{R\text{-gr}}(A) \to \mathrm{Aut}_{R\text{-gr}}(A) \xrightarrow{\varphi} \mathrm{PC}(R) = \mathrm{Pic}(R) \times G(R)$. The map φ is given by $I_f = \varphi(f) = \{u \in A : f(a)u = ua, \text{ for all } a \in A\}$, and $\mathrm{Im}(\varphi) = \{[I] \in \mathrm{PC}(R) : A \otimes I \cong_g A$ as left graded A-modules$\}$.

Proof: This is a very obvious adaption of [37, IV.1.2]. ∎

Note that it follows from VI.1.23 that if a graded isomorphism is inner, then it is induced by a homogeneous element. Indeed, if f is inner, then $I_f = Ru_f$ for some $u_f \in U(A)$ with u_f homogeneous. Now let A be a central G-dimodule algebra. Then $G(R)$ embeds in $\mathrm{Aut}_{R\text{-gr}}(R)$ using the G-action on A, so we have a sequence of morphisms

$$\alpha_A : G(R) \longrightarrow \mathrm{Aut}_{R\text{-gr}}(R) \longrightarrow \mathrm{Pic}(R) \times G(R) \longrightarrow G(R).$$

We define $\beta_A : G(R) \to G(R) : \sigma \to \sigma(\alpha_A(\sigma))^{-1}$.

VI.1.24 Theorem If $G = C_n$, then β_A induces an epimorphism

$$\beta : \mathrm{BAz}(R, G) \longrightarrow \mathrm{Aut}\, G(R) : [A] \longrightarrow \beta_A$$

Proof: Our proof is a slight generalization of [78, 1.2] and [98, 4.4].

(1.24.1) $\beta_A \in \mathrm{Aut}\, G(R)$:

Recall that we have a dimodule isomorphism $F : A \# \bar{A} \to \mathrm{END}_R(A)$, given by $F(a \# \bar{b})(x) = a^{\sigma} x b$, for $\deg b = \sigma$. Clearly α_A and β_A are group homomorphisms. Suppose that $\beta_A(\sigma) = 1$, that is, $\sigma_A(\sigma) = \sigma$. Take $u \in I_\sigma = \varphi(\sigma)$; then $\deg u = \sigma$. Now for all $x \in A, F(1 \# u)(x) = {}^{\sigma}xu = ux = F(u \# 1)(x)$, so

$1 \# u = u \# 1$; hence $u \in R$. Therefore $\deg u = 1$, so $\sigma = 1$. Thus β_A is a monomorphism, and therefore an isomorphism because G is finite.

(1.24.2) G acts trivially on I_σ:

First suppose that R is local. Then $I_\sigma = Ru_\sigma$, where $\deg u_\sigma = \alpha_A(\sigma)$, and σ is induced by u_σ. Let $f(\sigma,\tau) = u_\sigma u_\tau u_{\sigma\tau}^{-1}$; then f is a cocycle in $Z^2(G, U(R))$, and f is abelian, because G is cyclic; also, we can choose u_σ in such a way that f is normalized, and

$$\begin{aligned} {}^\tau u_\sigma &= u_\tau u_\sigma u_\tau^{-1} \\ &= f(\tau,\sigma) f(\tau,\tau^{-1})^{-1} f(\tau\sigma,\tau^{-1}) u_\sigma \\ &= f(\sigma\tau,\tau^{-1}) f(\sigma,1)^{-1} f(\sigma,\tau) f(\tau,\tau^{-1})^{-1} u_\sigma \\ &= u_\sigma \end{aligned}$$

by the cocycle relation. We used the fact that f is abelian.

(1.24.3) $\beta_{A\#B} = \beta_A \circ \beta_B$:

We have to show that, for every $\sigma \in G(R)$,

$$\begin{aligned} \sigma(\alpha_{A\#B}(\sigma))^{-1} = \beta_{A\#B}(\sigma) &= \beta_A(\beta_B(\sigma)) \\ &= \beta_B(\sigma)(\alpha_A(\beta_B(\sigma)))^{-1} \\ &= \sigma\alpha_B(\sigma)^{-1}(\alpha_A(\beta_B(\sigma)))^{-1} \end{aligned}$$

or

$$\alpha_{A\#B}(\sigma) = \alpha_A(\beta_B(\sigma))\alpha_B(\sigma)$$

Let $\alpha_B(\sigma) = \tau, \beta_B(\sigma) = \sigma\tau^{-1} = \rho, \alpha_A(\beta_B(\sigma)) = \gamma$. Take $u \in I_\rho, v \in J_\sigma = \varphi(B)(\sigma)$; then $\deg u = \gamma, \deg v = \tau$. For all $a \in A, b \in B$, we have

$$\begin{aligned} {}^\sigma(a \# b)(u \# v) &= ({}^\sigma a \# {}^\sigma b)(u \# v) \\ &= {}^\sigma a u \# {}^\sigma b v \qquad (G \text{ acts trivially on } u) \\ &= u {}^\tau a \# vb \\ &= (u \# v)(a \# b) \end{aligned}$$

Thus $u \# v \in \varphi_{A\#B}(\sigma)$, and $\alpha_{A\#B}(\sigma) = \deg(u \# v) = \tau\gamma = \alpha_A(\beta_B(\sigma))\alpha_B(\sigma)$.

(1.24.4) If $A = \mathcal{END}_R(P)$, then $\alpha_A(\sigma) = 1$, and $\beta_A = 1_G$:
Indeed, if P is a G-dimodule, then G acts on $\mathcal{END}_R(P)$ as follows: ${}^{\sigma}f(x) = {}^{\sigma}f({}^{\sigma^{-1}}x)$, so the action of G on f is induced by the automorphism $x \to {}^{\sigma}x$ on P, which is of degree 1. So $I_\sigma = 1$ in $\mathrm{PC}(R,G)$.

(1.24.5) β is surjective.
Take $j \in \mathrm{Aut}(G(R))$, and let $k : G \to G$ be defined by $k(\sigma) = \sigma j(\sigma)^{-1}$. Let $\mathrm{RG}(j)$ be the graded R-module RG, with action defined by the formula

$$ {}^{\sigma}u_\tau = u_{k(\sigma)\tau} $$

Observe that $\mathrm{RG}(j)$ is not a dimodule, because σ is an automorphism of degree $k(\sigma)$. However, $A = \mathrm{End}\,\mathrm{RG}(j)$, with the induced action and gradation, is a dimodule algebra, and clearly $\sigma_A = k, \beta_A = j$. It remains to be shown that A is a G-Azumaya algebra. It suffices to show that $F : A \# \bar{A} \to \mathcal{END}_R(A)$ is surjective (from a dimension argument, it follows that F is injective too, and a similar argument shows that $\bar{A} \# A \cong_d (\mathcal{END}_R(A))^{\mathrm{opp}}$). We know that $A \otimes A^{\mathrm{opp}} \cong \mathrm{End}_R(A)$. Take $f \in \mathrm{End}_R(A)$, and let $\Sigma_i f_i \otimes g_i$ be its preimage in $A \otimes A^{\mathrm{opp}}$. Suppose that $\deg g_i = \gamma_i$ and let $\rho_i = j^{-1}(\gamma_i)$. Then $\deg u_{k(\rho_i)} g_i = k(\rho_i)\gamma_i = \rho_i j(j^{-1}(\gamma_i))^{-1}\gamma_i = \rho_i$. Therefore, for all $h \in A$,

$$ \begin{aligned} F(\Sigma_i f_i u_{k(\rho_i)}^{-1} \# u_{k(\rho_i)} g_i)(h) &= \Sigma_i f_i u_{k(\rho_i)}^{-1}\, {}^{\rho_i}h u_{k(\rho_i)} g_i \\ &= \Sigma_i f_i h g_i \\ &= f(h) \quad \blacksquare \end{aligned} $$

VI.1.25 Lemma Suppose that $G = C_n$; then $\mathrm{BC}(R,G)$ and $\mathrm{BM}(R,G)$ are subgroups of $\mathrm{BD}^s(R,G)$.

Proof: Take $[A] \in \mathrm{BM}(R,G)$. Let σ be a generator for G, and take an étale covering S_1 of R splitting $\underline{A}$ in $\mathrm{Br}(R)$. Then $A \otimes S_1 \in \mathrm{End}_{S_1}(P_1)$. Take an étale covering S_2 of S_1 such that σ acts innerly on $A \otimes S_2$. We have that ${}^{\sigma}a = sas^{-1}$, and clearly $s^n \in Z(A \otimes S_2) =$

S_2. Take a faithfully flat S_2-algebra S containing an nth root ω of s^n. Replacing s by $s\omega^{-1}$, we may assume that $s^n = 1$. We define an action of G on $P = P_1 \otimes S$ as follows: ${}^\sigma x = sx$. This action induces the action of G on $A \otimes S$: for $a \in A \otimes S = \mathrm{END}_S(P)$, we have that $({}^\sigma a)x = {}^\sigma(a^{\sigma^{-1}}x) = sas^{-1}x$. Therefore $A \otimes S = \mathrm{End}_S(P)$, proving that $\mathrm{BM}(R,G) \subset \mathrm{BD}^s(R,G)$. The first assertion follows if we observe that $\mathrm{BC}(R,G) = \mathrm{BM}(R,G^*)$. Indeed, there is a one-to-one correspondence between the category of G-graded modules and G^*-modules: given a G-graded module M, we define an action of G^* on M by $\sigma^*(a) = \sigma^*(\deg a)a$, and conversely, if M is a G^*-module, then we define a G-gradation on M by

$$\deg a = \tau \text{ if and only if } \sigma^*(a) = \sigma^*(\tau)a, \text{ for all } \sigma^* \in G \quad \blacksquare$$

VI.1.26 Remark If G is noncyclic, then the foregoing lemma does not hold. For instance, take $G = C_2 \times C_2 = \langle\sigma\rangle \times \langle\tau\rangle$, and let k be a separably closed field of characteristic different from 2. Define a (nonabelian) normalized cocycle $f \in Z^2(G,U(k))$ as follows:

$$f(\sigma,\sigma) = f(\sigma,\tau) = f(\sigma\tau,\sigma) = f(\sigma\tau,\tau) = -1.$$

$$f(\alpha,\beta) = 1 \qquad \text{in all other situations.}$$

The twisted groupring kG_f represents a nontrivial element of $\mathrm{BC}(k,G)$. But from VI.1.19, it follows that $\mathrm{BC}^s(k,G) = 1$.

VI.1.27 Proposition Let $G = C_n$; then we have an exact sequence

$$1 \longrightarrow \mathrm{BD}^s(R,G) \longrightarrow \mathrm{BAz}(R,G) \xrightarrow{\beta} \mathrm{Aut}(G(R)) \longrightarrow 1$$

Proof: It is clear that $\mathrm{BD}^s(R,G) \subset \mathrm{Ker}(\beta)$. Indeed, take a flat covering S splitting $[A] \in \mathrm{BD}^s(R,G)$, and consider the commutative diagram

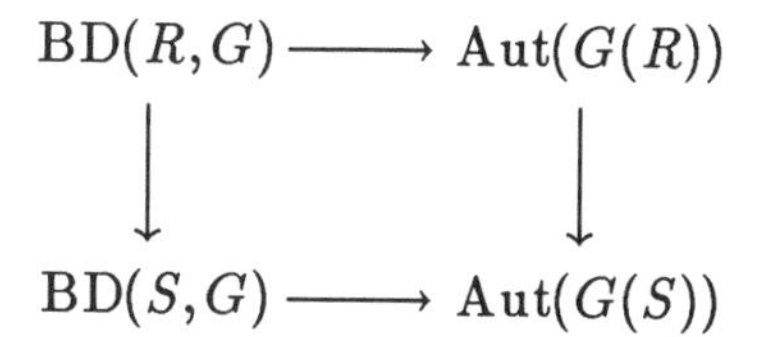

The result follows from the fact that $\mathrm{Aut}(G(R)) \to \mathrm{Aut}(G(S))$ is monomorphic. Take $[A] \in \mathrm{Ker}(\beta)$; then we can find a flat covering S of R such that $G(S)$ acts innerly on $A \otimes S$, the action being induced by elements of degree one, and such that $A \otimes S = \mathrm{End}_S(P)$. We therefore have that the action of $G(S)$ on $A \otimes S$ is induced by an action of $G(S)$ on P, so $[A \otimes S] \in \mathrm{BC}(S,G) \subset \mathrm{BD}^s(S,G)$. Thus $[A] \in \mathrm{BD}^s(R,G)$. ∎

Note It follows that $\mathrm{BAz}(R,G)$ is a product of $\mathrm{Aut}(G(R))$ and $\mathrm{BD}^s(R,G)$. The multiplication rules on this product are to appear in [82].

VI.1.28 Proposition Let $G = C_p$, where p is a prime number, and let R be a connected commutative ring.
(1.28.1) If p is not invertible in R, then $\mathrm{BD}(R,G) = \mathrm{BAz}(R,G)$.
(1.28.2) If p is invertible in R, then we have an exact sequence

$$1 \longrightarrow \mathrm{BAz}(R,G) \longrightarrow \mathrm{BD}(R,G) \longrightarrow C_2 \longrightarrow 1$$

where the nontrivial element of C_2 is represented by RG.

Proof: (1.28.1) follows from [98, 2.7]; (1.2.2) holds in the case when R is a separably closed field ([96, 2.5–2.7]), and actually the proof given there can be easily generalized to the case of a strictly henselian ring. Now let R be an arbitrary commutative ring, and suppose that A is a noncentral G-Azumaya algebra. The usual geometric argument shows that there is an étale covering S of R such that $A \otimes S \cong_d B \# SG$, where B is a central S-G-Azumaya algebra. Now $(A \# \overline{RG}) \otimes S \cong_d B \# SG \# \overline{SG}$ is S-central, so $A \# \overline{RG}$ is R-central, so $A \# \overline{RG}$ is R-central. Therefore $[A] = [A \# \overline{RG}][\mathrm{RG}] \in \mathrm{BAz}(R,G)[\mathrm{RG}]$. ■

The Brauer–Long group is now completely determined, up to the exact multiplication rule, in the exact sequence above. This is a rather complicated computation, and it has been carried out by DeMeyer and Ford in the case where $G = C_2$; we refer to [90].

For a noncyclic abelian group G, calculations become very cumbersome. Some partial results have been obtained by Beattie (cf. [81]), Deegan (cf. [89]), and Childs (cf. [86]); they involve the case where R is a separably closed field, and $G = C_p \times C_p$. Recently, the $\mathbf{Z}$-graded Brauer–Long group $\mathrm{BD}_g(R,G)$ of a $\mathbf{Z}$-graded ring R has been introduced by Tilborghs (cf. [103]). The special case of the $\mathbf{Z}$-graded Brauer–Wall group was studied by Tilborghs and Van Oystaeyen in [104]. Finally, let us mention that Long also has defined a Brauer group with respect to a Hopf algebra; this group generalizes the Brauer–Long group studied in this section. We refer to [97].

VI.2 THE BRAUER–WALL GROUP

VI.2.1 Definition Let R be a commutative ring, and suppose that R is connected (although this is not really necessary). Let G be the cyclic group of order 2, that is, $G = C_2 = \{1,-1\} = \{1,\sigma\}$. Consider the bilinear map $\varphi : G \times G \to U(R)$, defined by $\varphi(-1,-1) = -1$, and $\varphi(i,j) = 1$ in all other cases. The group $B_\varphi(R,C_2)$ is also denoted $\mathrm{BW}(R)$, and is called the *Brauer–Wall group.* Thus the Brauer–Wall group is the subgroup of the Brauer–Long group $\mathrm{BD}(R,C_2)$ consisting of classes of G-Azumaya algebras such that the action is defined by the gradation in the following way:

$${}^{\sigma}(a_1 + a_{-1}) = a_1 - a_{-1}$$

In the case of a field, $BW(R)$ was introduced by Wall in [107]; for the theory over a commutative ring, we refer to [101, 102].

In this section, we give a complete cohomological description of $BW(R)$. The results may be generalized to the Brauer group of Childs, Garfinkel, and Orzech for cyclic groups, but we shall restrict our attention to the Brauer–Wall group, because of its importance in the study of quadratic forms.

A G-Azumaya algebra representing an element of $BW(R)$ will be called an AZUMAYA algebra (following [101]); it is well known (cf. [101, 6.1]) that A is an AZUMAYA algebra if and only if A is finitely generated as an R-module, R-separable, and $\mathrm{CENTER}(A) = \{a \in A : ax = \varphi(\deg a, \deg x)xa, \text{ for all } x \in (A)\} = R$.

VI.2.2 The group of quadratic extensions If $2^{-1} \in R$, then is easily established that G and G^* are isomorphic sheaves on $R_{\text{ét}}$. Therefore, Galois extensions are also Galois* extensions, and we shall call them *quadratic extensions.* So quadratic extensions are C_2-dimodules, and the action is defined by the grading in the way described above. Also $G \cong \mu_2$ on $R_{\text{ét}}$, and we have an exact sequence of sheaves

$$1 \longrightarrow \mu_2 \longrightarrow U \longrightarrow U \longrightarrow 1$$

resulting in an exact sequence

$$1 \longrightarrow \mu_2(R) \longrightarrow U(R) \longrightarrow U(R) \longrightarrow H^1(R_{\text{ét}}, \mu_2)$$
$$\longrightarrow H^1(R_{\text{ét}}, U) \longrightarrow H^1(R_{\text{ét}}, U) \longrightarrow \cdots$$

As $H^1(R_{\text{ét}}, \mu_2) = H^1(R_{\text{ét}}, G) = T(R, G)$ and $H^1(R_{\text{ét}}, U) = \mathrm{Pic}(R)$, we have the following exact sequence:

$$1 \longrightarrow U(R)/U(R)^2 \longrightarrow T(R, G) \longrightarrow \mathrm{Pic}_2(R) \longrightarrow 1$$

The maps are defined as follows: for $a \in U(R)$, a quadratic extension $R\langle a\rangle$ is defined by $R\langle a\rangle = R[X]/(X^2 - a)$, where $\deg X = -1$, and for any quadratic extension S of R, an element of $\mathrm{Pic}_2(R)$ is given by S_{-1}. The neutral element of $T(R,G)$ is given by $R\langle 1\rangle = R \oplus Ru = Re_1 \oplus Re_2$, where $\deg u = -1, u^2 = 1, e_1, e_2$ are orthogonal idempotents, and ${}^{\sigma}e_1 = e_2, {}^{\sigma}e_2 = e_1$. The image of $R\langle a\rangle$ under the isomorphism $\beta : T(R,G) \to H^1(R_{\text{ét}}, G)$ will be denoted by c_a, and c_a is represented by the cocycle $-\sqrt{a} \otimes \sqrt{a}^{-1} \in Z^1(\underline{R\langle a\rangle}/R, \mu_2)$, or by $(1 \to 1, -1 \to -1)$ as an element of $Z^1(G,G)$ (note that $\mathrm{Gal}(R\langle a\rangle / R) = G$).

VI.2.3 Proposition Let R be a strictly henselian ring such that $2^{-1} \in R$, and let A be an AZUMAYA algebra. We have two possibilities:
(2.3.1) If A is R-central, then $A = M_n(R)(\underline{d})$ for some $\underline{d} \in G^n$.
(2.3.2) If A is not R-central, then $A = M_n(R)e_1 \oplus M_n(R)e_2 = M_n(R) \otimes R\langle 1\rangle$. Consequently $\mathrm{BW}(R) = C_2$.

Proof: First, suppose that A is central over R. Then $A \cong M_n(R)$. The action of σ on A is an automorphism, which is inner, so there exists $s \in A$ such that ${}^{\sigma}a = sas^{-1}$. Clearly $s^2 \in Z(A) = R$, so we may suppose that $s^2 = 1$ since R is strictly henselian. On R^n we define an action by ${}^{\sigma}x = sx$. On $M_n(R)$, this defines an action as follows: ${}^{\sigma}a(x) = {}^{\sigma}(a({}^{\sigma^{-1}}x)) = sas^{-1}x$, which is the original action of G on A. The action defines a gradation, and the result follows.

Suppose that A is not central over R. We have that $R \subset Z(A)^G$. Conversely, if $e = e_{-1} + e_1 \in Z(A)^G$, then $e = {}^{\sigma}e = -e_{-1} + e_1$, so $e = e_1$, and then $e \in \mathrm{CENTER}(A) = R$. So $R = Z(A)^G$. Now, suppose that R is a field, that is, R is separably closed. Then $J(A)$ is a graded ideal of A, because it is fixed by G. As $J(A) \# \bar{A}$ is a two-sided ideal of the simple algebra $A \# \bar{A} = \mathrm{END}_R(A)$, $J(A) \# \bar{A} = 0$, so A is simple, and $A = M_{n_1}(R)e_1 \oplus$

$\cdots \oplus M_{n_k}(R)e_k$. The e_i are central orthogonal idempotents, and $\sigma_{e_i} = e_j$. Since $R = Z(A)^G$, $k = 2$ and $n_1 = n_2$, so $A = M_{n_1}(R) \oplus M_{n_1}(R)e_2$. For a strictly henselian ring, the result follows from the lifting property for idempotents. ■

VI.2.4 Corollary Let R be a connected commutative ring such that $2^{-1} \in R$. If A is a central R-AZUMAYA algebra, then we may find an étale covering S of R, and $P \in \underline{\underline{FPC}}(S,G)$, such that $A \otimes S \cong_d \mathrm{END}_S(P)$. In this case, we call A a *+algebra.*

If A is not central, then we may find an étale covering S of R, and a (trivially graded) $P \in \underline{\underline{FP}}(S)$, such that $A \otimes S \cong_d \mathrm{End}_S(P) \#_S S\langle 1\rangle$. Such an algebra is called a *−algebra.*

Proof: An easy exercise. ■

The subgroup of $\mathrm{BW}(R)$ consisting of + algebras is called the *little Brauer–Wall group*; in the literature it is denoted by $\mathrm{BW}^+(R)$. From VI.2.4, it follows immediately that $\mathrm{BW}^+(R) = \mathrm{BW}^s(R)$. Then VI.1.20 provides a cohomological description of $\mathrm{BW}^+(R)$:

VI.2.5 Theorem We have an isomorphism $\eta : \mathrm{BW}^+(R) \rightarrow H^1(R_{\text{ét}}, G) \times_\psi H^2(R_{\text{ét}}, U)_{\text{tors}}$, where the multiplication is given by $(\sigma, u)(\tau, v) = (\sigma\tau, uv\psi(\sigma,\tau))$, with $\psi(\sigma,\tau) = \varphi(\tau_1, \tau_3)$.

In Sec. VI.1, we have seen that we have Chase–Rosenberg exact sequences for the Brauer–Long group and its subgroups. In particular, we have a map

$$\alpha : H^1(S/R,G) \times_\psi H^2(S/R,U) \longrightarrow \mathrm{BW}(S/R)$$

If $\alpha(\lambda,u) = [A]$, then $\eta([A])$ is the image of (λ,u) in $H^1(R,G) \times_\psi H^2(R,U)_{\text{tors}}$. If S is a Galois extension of R with group G_1, then we obtain a map

$$\alpha : H^1(G_1,G) \times_\psi H^2(G_1,U(S)) \longrightarrow \mathrm{BW}(S/R)$$

We give a description of this map. Take $(\lambda, u) \in H^1(G_1, G) \times_\psi H^2(G, U(S))$. As an R-algebra, $A = \alpha(\lambda, u)$ is given by

$$A = \oplus_{\sigma \in G_1} S u_\sigma$$

with the multiplication rules induced by the cocycle u and the action of G_1 on S. The G-gradation on A is determined by $\deg(u_\sigma) = \lambda(\sigma)$. We refer to III.5.10 for the analog in the $\mathbf{Z}$-graded case.

If 2 is not invertible in R, then $\mathrm{BW}^s(R) = \mathrm{BW}(R)$, by VI.1.29, and VI.2.5 describes $\mathrm{BW}(R)$ completely; from now on, we suppose that $2^{-1} \in R$.

We return to quadratic extensions. Obviously a quadratic extension L of R is an R-AZUMAYA algebra. Indeed, $L \# \underline{L} \cong_d \underline{L}\langle 1 \rangle$ is an $\underline{L}$-Azumaya algebra. Clearly L is a $-$algebra, and if we take the smash product of two quadratic extensions, then we obtain a $+$algebra.

VI.2.6 Proposition Let L, M be two quadratic extensions of R, corresponding respectively to $\lambda, \mu \in H^1(R_{\text{ét}}, G)$. Then $\eta([L \# M]) = (\lambda\mu c_{-1}, \psi(\lambda, \mu))$.

Proof: First we consider the case $M = R\langle 1 \rangle$. From VI.2.2, we know that $L = R \oplus I$, where $I \in \mathrm{Pic}_2(R)$, and where the multiplication on L is defined by a fixed isomorphism $f : I \oplus I \to R$. Denote $\mathrm{Gal}(L/R) = \{1, \sigma\}$; then $\lambda \in H^1(R_{\text{ét}}, G)$ is represented by the cocycle $(1 \to 1, \sigma \to -1)$ in $H^1(\{1, \sigma\}, G)$. This cocycle is still denoted by λ.

Similarly, let $C = R\langle -1 \rangle$ and $\mathrm{Gal}(C/R) = \{1, \tau\}$. Let $S = C \otimes \underline{M}$; then $\mathrm{Gal}(S/R) = \{1, \sigma, \tau, \sigma\tau\}$, and λc_{-1} is represented by the cocycle $(1 \to 1, \sigma\tau \to 1, \sigma \to -1, \tau \to -1)$. We calculate the image $[A]$ of $(\lambda c_{-1}, 1)$ in $\mathrm{BW}(S/R)$:

$$A = S \oplus S u_\sigma \oplus S u_\tau \oplus S u_\sigma u_\tau$$

where

$$u_\sigma^2 = u_\tau^2 = 1 \qquad u_\sigma u_\tau = u_\tau u_\sigma \qquad \deg u_\sigma = \deg u_\tau = -1$$

$$x u_\sigma = -u_\sigma x \qquad x u_\tau = u_\tau x \qquad \text{for } x \in I$$

$$i u_\sigma = u_\sigma i \qquad i u_\tau = -u_\tau i$$

where $C = R \oplus Ri, i^2 = -1$. As an R-algebra, A is generated by

$$u_\tau, \quad iu_\sigma, \quad iu_\tau, \quad iu_\sigma I$$

These elements anticommute, and they all have degree -1. Furthermore, $v^2 = (iu)^2 = 1$ and $(iv)^2 = -1$, and f induces an isomorphism $iu_\sigma I \otimes iu_\sigma I \to R$. Therefore $A \cong_d R\langle -1\rangle \# R\langle 1\rangle \# R\langle 1\rangle \# L$. Now, we calculate $R\langle -1\rangle \# R\langle 1\rangle$. Write $R\langle -1\rangle = R \oplus Ri$, $R\langle 1\rangle = R \oplus Ru$. We have an isomorphism $f : R\langle -1\rangle \# R\langle 1\rangle \to M_2(R\langle -1\rangle)(-1,1)$ given by

$$f(1 \# 1) = \begin{bmatrix} 1 & 0 \\ 0 & 1 \end{bmatrix} \qquad f(i \# 1) = \begin{bmatrix} 0 & i \\ i & 0 \end{bmatrix}$$

$$f(1 \# u) = \begin{bmatrix} 0 & 1 \\ 1 & 0 \end{bmatrix} \qquad f(i \# u) = \begin{bmatrix} i & 0 \\ 0 & i \end{bmatrix}$$

f is an isomorphism of dimodule algebras, so $R\langle -1\rangle \# R\langle 1\rangle$ represents the trivial element of $\mathrm{BW}(R)$. So we have that $[A] = [R\langle 1\rangle \# L]$ in $\mathrm{BW}(R)$, and we have shown that $\eta(R\langle 1\rangle \# L) = (\lambda c_{-1}, 1)$. It follows that $\eta(R\langle 1\rangle \# R\langle 1\rangle) = (c_{-1}, A)$, and $\eta(\overline{R\langle 1\rangle \# R\langle 1\rangle}) = (c_{-1}, 1)^{-1} = (c_{-1}, \psi(c_{-1}, c_{-1}))$. In general,

$$\begin{aligned} \eta([L \# M]) &= \eta([L \# M][R\langle 1\rangle \# R\langle 1\rangle][\overline{R\langle 1\rangle \# R\langle 1\rangle}]) \\ &= \eta([L \# R\langle 1\rangle])\eta([M \# R\langle 1\rangle])\eta([\overline{R\langle 1\rangle \# R\langle 1\rangle}]) \\ &= (\lambda c_{-1}, 1)(\mu c_{-1}, 1)(c_{-1}, \psi(c_{-1}, c_{-1})) \\ &= (\lambda\mu c_{-1}, \psi(\lambda, \mu)) \end{aligned}$$

where we use the fact that ψ is bilinear. ■

VI.2.7 Theorem We have an isomorphism of groups

$$\theta : \mathrm{BW}(R) \longrightarrow X = \{+,-\} \times H^1(R_{\text{ét}}, G) \times H^2(R_{\text{ét}}, U)_{\text{tors}}$$

where the multiplication is given by

$$(+,\lambda,u)(+,\mu,v) = (+,\lambda\mu,uv\psi(\lambda,\mu))$$
$$(-,\lambda,u)(-,\mu,v) = (+,\lambda\mu c_{-1},uv\psi(\lambda,\mu))$$
$$(+,\lambda,u)(-,\mu,v) = (-,\lambda\mu,uv\psi(\lambda,\mu c_{-1}))$$

Proof: From the preceding sections, we know that we have isomorphisms $\alpha : \mathrm{Br}(R) \to H^2(R_{\text{ét}}, U)_{\text{tors}}$, $\beta : T(R,G) \to H^1(R_{\text{ét}}, G)$. Take $[A] \in \mathrm{BW}(R)$. If A is central, then we define $\theta([A]) = (+,\eta([A]))$, and this defines a monomorphism $\mathrm{BW}^+(R) \to X$. If A is not R-central, then there exists an étale covering S of R such that $A \otimes S \cong_d \mathrm{End}_S(P) \# S\langle 1\rangle$, where P is a trivially graded faithfully projective S-module. Restricting to the part of degree 1, we obtain an isomorphism $A_1 \otimes S \cong \mathrm{End}_S(P)$, and restricting to the center, we obtain that $Z(A) \otimes S \cong S\langle 1\rangle$. It follows that A_1 is R-Azumaya, and that $Z(A)$ is a quadratic extension of R. We define $\theta([A]) = (-,\beta(Z(A)),\alpha(A_1))$. Let us check that θ is a homomorphism. Take $L, M \in T(R,G)$, and $A_1, B_1 \in \mathrm{Br}(R)$. Let $\alpha([A_1]) = u, \alpha([B_1]) = v, \beta(L) = \lambda, \beta(M) = \mu$, and suppose that A, B are R-Azumaya algebras such that $\theta([A]) = (-,\lambda,u)$ and $\beta([B]) = (-,\mu,v)$. Then

$$\begin{aligned}\theta([A \# B]) &= \theta([A_1 \# B_1 \# L \# M]) \\ &= \theta([A_1])\theta([B_1])\theta([L \# M]) \\ &= (+,1,u)(+,1,v)(+,\lambda\mu c_{-1}\psi(\lambda,\mu)) \\ &= (+,\lambda\mu c_{-1}, uv\psi(\lambda,\mu))\end{aligned}$$

Finally, suppose that A and B are AZUMAYA algebras such that $\theta([A]) = (+,\lambda,u)$ and $\theta([B]) = (-,\mu,v)$. Choose Azumaya algebras A_1 and B_1 such that $\alpha([A_1]) = u\psi(\lambda\mu,\mu c_{-1})$ and $\alpha([B_1]) = v$, and quadratic extensions L, M, N such that

$\beta(L) = \mu c_{-1}, \beta(M) = \mu$ and $\beta(N) = \lambda\mu$. Then

$$\theta([A_1 \# L \# N]) = (+, \lambda, u) = \theta([A])$$
$$\theta([B_1 \# M]) = (-, \mu, v) = \theta([B])$$

Therefore

$$\theta([A \# B]) = \theta([A_1 \# L \# N \# B_1 \# M])$$

we have that

$$\theta([A_1 \# B_1]) = (+, 1, uv\psi(\lambda\mu, \mu c_{-1}))$$
$$\theta([M \# L]) = (+, 1, \psi(\mu, \mu c_{-1}))$$
$$\theta([N]) = (-, \lambda\mu, 1)$$

so

$$\theta([A \# B]) = (-, \lambda\mu, uv\psi(\lambda, \mu c_{-1})) \quad \blacksquare$$

VI.3 GRADED BRAUER GROUPS IN A GEOMETRICAL CONTEXT

In this section we provide a description of the Brauer group of a projective variety or scheme in terms of some Brauer groups. For technical reasons we restrict our attention to low-dimensional cases, that is, dimension one and two. The more general situation requires the theory of relative Brauer groups in the sense of [69] and this is outside the scope of this work.

VI.3.1 Let R be a noetherian integrally closed domain; when considering properties of a graded nature we assume that R has a **Z**-gradation. Furthermore, we say that R is *regular in codimension* n if for all prime ideals p of R of height less then n, the ring R_p is a regular local ring. Write $X^{(n)}(R)$ for $\{p \in \mathrm{Spec}(R) : \mathrm{ht}(p) \leq n\}$; we obtain a filtration

(3.1.1) $$0 \subset X^{(1)}(R) \subset \cdots \subset X^{(n)}(R) \subset \cdots \subset X^{(d)}(R) = \mathrm{Spec}(R)$$

where d = Krull dim(R). We will write $X^i(R)$ for $X^{(i)}(R) \setminus X^{(i-1)}(R)$ and clearly any Zariski open set containing $X^i(R)$ also contains $X^{(i)}(R)$. For any n such that R is regular in codimension n we define $\mathrm{Br}^{(n)}(R) = \cap\{\mathrm{Br}(R_p) : p \in X^{(n)}(R)\}$. Since any noetherian integrally closed domain is necessarily regular in codimension 1, we may always define $\beta(R) = \mathrm{Br}^{(1)}(R) = \cap\{\mathrm{Br}(R_p) : p \in X^{(1)}(R)\}$.

First note that these definitions make sense; indeed, for the regular local rings R_p we know that $\mathrm{Br}(R_p) \subset \mathrm{Br}(K)$, where K is the field of fractions of R, so the intersection of the groups listed is understood to be taken in $\mathrm{Br}(K)$. It is obvious that $\mathrm{Br}^{(d)}(R) = \mathrm{Br}(R)$ if R is regular of dimension d, because an R-algebra A representing an element of $\mathrm{Br}^{(d)}(R)$ has the property that A_p is an R_p-Azumaya algebra for all prime ideals p of R; hence A is R-Azumaya by the local characterization of Azumaya algebras. The filtration (3.1.1) now corresponds to a filtration

$$(3.1.2)\quad \mathrm{Br}(R) = \mathrm{Br}^{(d)}(R) \subset \mathrm{Br}^{(d-1)}(R) \subset \cdots \subset \mathrm{Br}^{(n)}(R) \subset \cdots \subset \beta(R) \subset \mathrm{Br}(K)$$

in case R is a regular domain. If R is regular in codimension n with $n < d$ then we obtain only the part

$$(3.1.3)\quad \mathrm{Br}^{(n)}(R) \subset \cdots \subset \beta(R) \subset \mathrm{Br}(K)$$

Remark If R is not regular in codimension n, then we can still define $\mathrm{Br}^{(n)}(R)$. We then have to use the theory of relative Brauer groups (cf. [69]), and we obtain a sequence of morphisms:

$$\mathrm{Br}(R) = \mathrm{Br}^{(d)}(R) \longrightarrow \mathrm{Br}^{(d-1)}(R) \longrightarrow \cdots \longrightarrow \mathrm{Br}^{(n)}(R) \subset \mathrm{Br}^{(n-1)}(R) \subset \cdots \subset \beta(R) \subset \mathrm{Br}(K)$$

VI.3.2 Lemma Let R be regular in codimension n and let A be an R-algebra representing an element of $\mathrm{Br}^{(n)}(R)$. Then A

is a maximal R-order in its ring of quotients $\Sigma = Q(A)$ which represents the image of $[A]$ in $\mathrm{Br}(K)$.

Proof: If $\alpha = [A] \in \mathrm{Br}^{(n)}(R)$ then there is a $\Sigma \in \mathrm{Br}(K)$ such that for every $p \in X^{(n)}(R)$ there is an R_p-Azumaya algebra $A(p)$ such that $K \otimes_{R_p} A(p) = Q(A(p)) = \Sigma$. We identify the $A(p)$ with their canonical image in Σ and then we put $A = \cap\{A(p) : p \in X^{(n)}(R)\}$. It is very easy to verify that $Z(A) = R$ and therefore A is a prime polynomial identity (P.I.) algebra. Take $z \in A$ and consider its minimal polynomial over K. Since $z \in A_p$ and A_p is a finitely generated R_p-module for all $p \in X^{(n)}(R)$, z is integral over R_p for all $p \in X^{(n)}(R)$. But then the minimal polynomial of z over K has its coefficients in each R_p, for all $p \in X^{(n)}(R)$. Now since R is a Krull domain we know that $R = \cap\{R_p : p \in X^{(1)}(R)\} = \cdots = \cap\{R_p : p \in X^{(n)}(R)\}$, and therefore that the minimal polynomial of z has its coefficients in R. Consequently A is integral over R, hence an R-order. In fact it follows readily from a general result of G. Cauchon on prime P.I. rings with noetherian center that A is finitely generated as an R-module! Now at each $p \in X^{(n)}(R)$ (in fact it suffices to use $p \in X^{(1)}(R)$ here), $R_p \otimes_R A = A_p$ is a maximal order over R_p and from this it follows easily that A is a maximal R-order in Σ. ■

The group $\mathrm{Br}^{(1)}(R) = \beta(R)$ is called the *reflexive Brauer group* of R; it has been studied by S. Yuan in [108]. An R-algebra A representing an $\alpha \in \beta(R)$ is called a *reflexive Azumaya algebra*; consequently, for a ring R that is regular in codimension n, every $\gamma \in \mathrm{Br}^{(m)}(R)$ with $m \leq n$ may be represented by a reflexive Azumaya algebra Γ having some extra good properties in primes of codimension between 1 and m. Let us recall an elementary exact sequence relating $\mathrm{Br}(R)$ and $\beta(R)$:

VI.3.3 Lemma There is an exact sequence

$$0 \longrightarrow \mathrm{Pic}(R) \longrightarrow \mathrm{Cl}(R) \longrightarrow \mathrm{BCl}(R) \longrightarrow \mathrm{Br}(R) \longrightarrow \beta(R),$$

where $\mathrm{BCl}(R)$, the so-called Brauer-classgroup, is defined by the exactness of the sequence.

Proof: Cf. S. Yuan [108]. ■

If $S \supset R$ is an integral extension of an integrally closed noetherian domain such that the corresponding extension $L \supset K$ of the fields of fractions is a Galois extension with group G, then we introduce the following groups: $\mathrm{Div}(R)$, $\mathrm{Div}(S)$ stand for the divisor groups of R and S, respectively, and $C(S/R)$ will denote the kernel of the canonical map $i : \mathrm{Div}(R) \to \mathrm{Div}(S)$; $\beta(S/R)$ is the kernel of the morphism $\beta(R) \to \beta(S)$, $[A] \to [B]$, where $B = \cap\{(S \otimes_R A)_q : q \in X^{(1)}(S)\}$. A result of D. S. Rim (cf. [100]) then yields:

VI.3.4 Proposition With notation as above, the following sequence is exact:

$$\begin{aligned} 0 &\longrightarrow C(S/R) \longrightarrow H^1(G,U(S)) \longrightarrow \mathrm{Div}(S)^G/i\,\mathrm{Div}(R) \\ &\longrightarrow \mathrm{Cl}(S)^G/i\,\mathrm{Cl}(R) \longrightarrow H^2(G,U(S)) \longrightarrow \beta(S/R) \\ &\longrightarrow H^1(G,\mathrm{Cl}(S)) \longrightarrow H^3(G,U(S)), \end{aligned}$$

where $\mathrm{Div}(S)^G$ and $\mathrm{Cl}(S)^G$ are the fixed groups for the canonical G-action defined on $\mathrm{Div}(S)$ and $\mathrm{Cl}(S)$, respectively.

If the reader is interested in obtaining the modified versions of Lemma VI.3.3 and Proposition VI.3.4 for the Brauer groups $\mathrm{Br}^{(n)}(R)$, then the $\mathrm{Cl}(R)$ and $\mathrm{Cl}(S)$ have to be replaced by $\mathrm{Cl}(R)_n$, $\mathrm{Cl}(S)_n$ as introduced by Fossum, Claborn [88]. However, let us point out that this can easily be obtained from the general theory of relative Brauer and Picard groups (cf. Van Oystaeyen, Verschoren [69], and Caenepeel, Verschoren [84]), and the exact sequences relating these groups. These relative invariants have been defined in [69] with respect to generically closed subsets of Spec(R) (or in fact even more abstract topological spaces). What it comes down to is the introduction of certain reflector functors, that is localizations. In the situation here we may view κ_n as the kernel functor obtained by taking the infimum of $\{\kappa_p : p \in X^{(n)}(R)\}$, $\mathrm{Br}^{(n)}(R) = \mathrm{Br}(R, \kappa_n)$, and $\mathrm{Cl}(R)_n = \mathrm{Pic}(R, \kappa_n)$ in the terminology of [69]. In particular : $\beta(R) = \mathrm{Br}(R, \kappa_1)$ and $\mathrm{Cl}(R) = \mathrm{Cl}(R)_1 = \mathrm{Pic}(R, \kappa_1)$. Let us provide just somewhat more detail in the case of $\beta(R)$, enough to shed some light on the connections with reflexive modules, but leaving the case $n > 1$ to the reader as a not completely trivial exercise.

Now we assume that R is **Z**-graded. The subset $X_g^{(n)}(R)$ of $X^{(n)}(R)$ consists of the graded prime ideals of height n. A priori there is a problem here if one considers the height of a graded prime ideal in $\mathrm{Spec}^g(R)$ or in Spec(R); that there is no ambiguity is shown by the following lemma.

VI.3.5 Lemma Let R be a noetherian commutative ring. If p is a graded prime ideal with $\mathrm{ht}(p) = n$, then there exists a chain of graded prime ideals

$$p_0 \subsetneq p_1 \subsetneq \cdots \subsetneq p_n = p$$

that is, $\text{gr-ht}(p) = n$.

Proof: We use induction on n. For $n = 0$, there is nothing to prove. So suppose that the lemma is true for $n - 1$. If $\mathrm{ht}(p) = n$, then there exist a chain of distinct prime ideals: $q_0 \subset q_1 \subset \cdots \subset q_n = p$. If q_{n-1} is graded, then the induction hypothesis may be applied to q_{n-1} and the statement follows. Suppose that q_{n-1} is not graded and consider $(q_{n-1})_g$, the graded ideal generated by the homogeneous elements of q_{n-1}. In general, we have that $\mathrm{ht}(q) = \mathrm{ht}(q_g)$ or $\mathrm{ht}(q) = \mathrm{ht}(q_g) + 1$; thus $\mathrm{ht}((q_{n-1})_g) = n - 2$. The induction hypothesis yields the existence of a chain of distinct graded prime ideals $p_0 \subset p_1 \subset \cdots p_{n-2} = (q_{n-1})_g$. Put $\bar{R} = R/p_{n-2}$, and let $\bar{p}, \bar{q}_{n-1}$ be the images of p, q_{n-1}, respectively, in $\bar{R}$. Take a nonzero homogeneous element $\bar{a}$ in $\bar{p}$, and let $\bar{p}_{n-1}$ be a prime ideal of height one containing $\bar{a}$ (apply Krull's principal ideal theorem). Then $\bar{p}_{n-1}$ is graded, because otherwise we would have a chain of distinct prime ideals $0 \subset (\bar{p}_{n-1})_g \subset \bar{p}_{n-1}$. Also, $\bar{p}_{n-1} \neq \bar{p}_n$ since $\mathrm{ht}(\bar{p}_n) = 2$. Now let p_{n-1} be the prime ideal of R such that $\bar{p}_{n-1} = p_{n-1}/p_{n-2}$; then we obtain a chain of distinct graded prime ideals

$$p_0 \subset p_1 \subset \cdots \subset p_{n-2} \subset p_{n-1} \subset p_n = p \quad \blacksquare$$

If R is regular in codimension n then we define $\mathrm{Br}_g^{(n)}(R) = \cap\{\mathrm{Br}_g(Q_p^g(R)) : p \in X_g^{(m)}(R)\}$.

VI.3.6 Lemma Let R be regular in codimension n and take $m \leq n$; then $\mathrm{Br}_g^{(m)}(R) \subset \mathrm{Br}^{(m)}(R)$. We obtain a commutative diagram of inclusions:

$$\begin{array}{ccccccc}
\mathrm{Br}^{(n)}(R) \subset \cdots \subset & \beta(R) = \mathrm{Br}^{(1)}(R) & & \subset & & \mathrm{Br}(K) \\
\cup & \cup & & & & \cup \\
\mathrm{Br}_g^{(n)}(R) \subset \cdots \subset & \beta_g(R) = \mathrm{Br}_g^{(1)}(R) & \subset & \mathrm{Br}_g(K^g) & \subset & \mathrm{Br}(K^g)
\end{array}$$

where K^g is the graded field of fractions of R. If $n = \text{Krull dim}(R)$ then we have $\mathrm{Br}^{(n)}(R) = \mathrm{Br}(R)$ and $\mathrm{Br}_g^{(n)}(R) = \mathrm{Br}_g(R)$

Proof: We have only to check that an R-algebra A representing an $\alpha \in \mathrm{Br}_g^{(m)}(R)$ also represents an element, again denoted α, of $\mathrm{Br}^{(m)}(R)$. This comes down to verifying that A_p is an Azumaya algebra over R_p for all $p \in X^{(m)}(R)$, knowing that this condition holds for those p in $X^{(m)}(R) \cap \mathrm{Spec}^g(R) = X_g^{(m)}(R)$ (cf. VI.3.5). If $p \in X^{(m)}(R)$ is such that $p = p_g$ then $p \in X_g^{(m)}(R)$ and A_p is an Azumaya algebra over R_p by assumption. If $p \in X^{(m)}(R)$ is such that $p \neq p_g$ then the set of homogeneous elements in $R \setminus p$, say $h(R \setminus p)$, equals the set of homogeneous elements in $R \setminus p_g$.

Hence A_p is a localization of $Q^g_{p_g}(A)$; the latter being an Azumaya algebra over $Q^g_{p_g}(R)$ because $\mathrm{ht}(p_g) = \mathrm{ht}(p) - 1$, it also follows that A_p is an Azumaya algebra over R_p. ■

VI.3.7 Proposition Let R be regular in codimension n and take $m \leq n$. If A is a **Z**-graded R-algebra representing an $\alpha \in \mathrm{Br}^{(m)}(R)$, then $\alpha \in \mathrm{Br}^{(m)}_g(R)$.

Proof: For a graded prime ideal $p \in X^{(m)}_g(R)$ we know that A_p is an Azumaya algebra over R_p and we have to prove that $B = Q^g_p(A)$ is an Azumaya algebra over $Q^g_p(R)$. Clearly B is a prime P.I. algebra, so it will suffice to establish that its Formanek center $F(B)$ equals $Z(B)$ by showing that there is a multilinear central polynomial f for B such that f assumes an invertible value in $Z(B) = Q^g_p(R)$. Since A_p is an Azumaya algebra there is a multilinear central polynomial g for A_p such that its set of evaluations, denoted by $g(A_p)$, is not contained in the maximal ideal pR_p. Since B and A_p have the same P.I. degree, the multilinear central polynomial $g_{|B} = f$ does not vanish on B and even $f(B) \not\subset pZ(B)$, because A_p is a central extension of B. Now B is graded; if all homogeneous substitutions in f yield values in $pZ(B)$, then by multilinearity of f, all substitutions in B would yield values in $pZ(B)$ and this is excluded. But all homogeneous elements in $Z(B) \backslash pZ(B)$ are invertible by construction, hence the multilinear central polynomial f for B takes an invertible value and therefore B is an Azumaya algebra. All assertions now follow easily. ■

The graded reflexive Brauer group $\beta_g(R)$ will appear as the Brauer group of projective regular curves and surfaces if R stands for the homogeneous coordinate ring. At least for curves we will provide some results in the singular case as well, using Mayer–Vietoris sequences and the normalization diagram. Before deriving these geometrical interpretations of $\beta_g(R)$ for low dimensional rings let us provide a general cohomological description of $\beta(R)$ in terms of the set of regular points X_{reg} on X, the normal affine variety with coordinate ring R. That X_{reg} has to be taken into account is evident from the following observations.

VI.3.8 Theorem (Hoobler [31]) If R is a regular affine domain over a field k, then we have that $\beta(R) \cong \text{Br}(R)$.

VI.3.9 Remark Consider, for example, $R = C[X,Y,Z]/(X^2 - YZ)$. Then $\text{Cl}(R) = \mathbf{Z}/2\mathbf{Z}$ and this affine cone is generated by a ruling P. The algebra $\Lambda = \text{END}_R(R \oplus P)$ is a reflexive Azumaya algebra but it is not Azumaya in the cone. Nevertheless in the case $\dim R \leq 2$ we have that a reflexive module over a regular local ring of dimension ≤ 2 is a free module, and hence a reflexive Azumaya algebra is Azumaya on the open set of regular points. In this case $\beta(R) = \text{Br}(X_{\text{reg}})$ follows easily. For the definition of $\text{Br}(X_{\text{reg}})$ we refer to the next section.

VI.3.10 Theorem (Le Bruyn [95]) Let R be a normal affine domain, the coordinate ring of a normal affine variety X over an algebraically closed field.

(3.10.1) $\beta(R) \cong H^2(X_{\text{reg,ét}}, U)$. Note that the schemes appearing are all regular; hence it follows that the cohomology groups, and thus $\beta(R)$, are torsion groups.

(3.10.2) $\beta(R) = \text{Br}(V)$ for some open subset V in X_{reg}.

(3.10.3) Let Z be the closed subvariety $Z = X \setminus X_{\text{reg}}$. Then there is a long exact sequence in étale cohomology (cf. [44]) that may be combined with the sequence of Lemma VI.3.3 to yield, by diagram chasing, some cohomological interpretation for the otherwise rather unclear Brauer-class group

$B\,\mathrm{Cl}(R)$:

$$\begin{array}{ccccccccccccc}
H^1_Z(X,U) & \to & H^1(X,U) & \to & H^1(X_{\mathrm{reg}},U) & \to & H^2_Z(X,U) & \to & H^2(X,U) & \to & H^2(X_{\mathrm{reg}},U) & \to & H^3_Z(X,U) \\
& & \uparrow \cong & & \uparrow \cong & & \uparrow & & \uparrow & & \uparrow & & \\
& & \mathrm{Pic}(R) & \to & \mathrm{Cl}(R) & \to & B\,\mathrm{Cl}(R) & \to & \mathrm{Br}(R) & \to & \beta(R) & &
\end{array}$$

where $H^i_Z(X,U)$ denote the cohomology groups with support on Z (as in Milne's book [44]); all cohomology is taken on the étale site.

Proof: (1) and (2): Since R is normal, X_{reg} is determined by an ideal I of R such that $\mathrm{ht}(I) \leq 2$. Therefore there exist $u_1, u_2 \in I$ such that $\mathrm{ht}(Ru_1 + Ru_2) = 2$. Let V be the open set determined by the ideal $Ru_1 + Ru_2$ and cover V by $U_i = X(Ru_i)$, $i = 1,2$. Since $\beta(R)$ is defined by the Brauer groups in the height one prime ideals, $\beta(R)$ is the pullback of the diagram

$$\begin{array}{ccc}
\beta(R) & \longrightarrow & \beta(R_{u_1}) \\
\downarrow & & \downarrow \\
\beta(R_{u_2}) & \longrightarrow & \beta(R_{u_1 u_2})
\end{array}$$

Up to replacing "affine" by "localization of affine" in the statement of Hoobler's theorem (check in the proof that this is allowed) we can obtain $\beta(R)$ as the pullback in the diagram

$$\begin{array}{ccc}
\beta(R) & \longrightarrow & \mathrm{Br}(U_1) \\
\downarrow & & \downarrow \\
\mathrm{Br}(U_2) & \longrightarrow & \mathrm{Br}(U_1 \cap U_2)
\end{array}$$

Since only regular schemes appear in this proof, it follows that $H^2(\cdot, U)$ is a torsion group because of the natural injection

$H^2(\cdot,U) \to H^2(Q(\cdot),U) = \mathrm{Br}(Q(\cdot))$. Using Gabber's theorem we finally arrive at the pullback diagram

$$\begin{array}{ccc} \beta(R) & \longrightarrow & H^2(U_1,U) \\ \downarrow & & \downarrow \\ H^2(U_2,U) & \longrightarrow & H^2(U_1 \cap U_2, U) \end{array}$$

It follows that $\beta(R) \cong H^2(V,U) \cong \mathrm{Br}(V)$, proving 2.

To obtain a complete proof for (1) we apply Grothendieck's result on the cohomological purity of the Brauer group (cf. [28, III.6; 44, VI.5]) to the situation $V \subset X_{\mathrm{reg}}$; we obtain $H^2(V,U) \cong H^2(X_{\mathrm{reg}},U)$.

The proof of (3) is clear from the long exact sequence for the étale cohomology ([44, III.1.25]), taking into account that $H^1(X_{\mathrm{reg}},U) \cong \mathrm{Cl}(R)$ because a regular local ring is factorial; moreover, $H^1(X,U)$ classifies locally invertible modules up to isomorphism, while $\mathrm{Br}(R) \cong H^2(X,U)_{\mathrm{tors}}$ by Gabber's result. ■

Let us now focus on the graded situation. First note that Lemma VI.3.2 and the proof of Lemma VI.3.6 yield that for a **Z**-graded ring R regular in codimension n, any $\alpha \in \mathrm{Br}_g^{(m)}(R)$ is represented by a **Z**-graded maximal R-order A for any $m \leq n$.

VI.3.11 Lemma Let R be a **Z**-graded noetherian integrally closed domain; then an R-algebra A represents an element $\alpha \in \beta_g(R)$ if and only if for all $p \in X_g^{(1)}(R)$, $Q_p^g(A)$ is an Azumaya algebra over $Q_p^g(R)$.

Proof: Since R is regular in codimension 1, we apply Lemma VI.3.6; we use Lemma VI.3.2 to represent α by a (graded) maximal order, and in fact the proof of VI.3.6 also yields a proof for the claim above. ■

VI.3.12 Lemma Let A be a graded reflexive Azumaya algebra over a $\mathbf{Z}$-graded noetherian integrally closed domain. If p is a graded prime ideal of R then $Q_p^g(A)$ is a graded reflexive Azumaya algebra over $Q_p^g(R)$.

Proof: Since A is a maximal R-order, $Q_p^g(A)$ is a maximal (graded) $Q_p^g(R)$-order. If $\bar{q}_1$ is a graded prime ideal of height one in $Q_p^g(R)$ then $\bar{q}_1 = Q_p^g(R)q_1$ where q_1 is a graded prime ideal of height one contained in p. Therefore $Q_{\bar{q}_1}(Q_p^g(A)) = Q_{q_1}(A)$ is an Azumaya algebra and Lemma VI.3.11 implies that $Q_p^g(A)$ is a graded reflexive Azumaya algebra. ∎

Recall also the following from M. Auslander, O. Goldman [76]:

VI.3.13 Lemma If S is a noetherian domain of global dimension at most 2, then every finitely generated reflexive S-module is projective. In particular, if R is a noetherian integrally closed domain of global dimension at most 2, then $\mathrm{Br}(R) = \beta(R)$.

VI.3.14 Theorem Let R be a $\mathbf{Z}$-graded noetherian integrally closed domain of gr-global dimension at most 3. Suppose that R_0 is a field, and that R is positively graded. If A represents an $\alpha \in \beta_g(R)$, then for all $p \in \mathrm{Proj}(R)$ we have that $Q_p^g(A)$ is an Azumaya algebra.

Proof: Since p is graded and in $\mathrm{Proj}(R)$ we have $p \subsetneq R_+$, because R_+ is the unique gr-maximal ideal (R_0 is a field). Therefore $Q_p^g(A)$ has dimension at most two. By Lemma VI.3.12, $Q_p^g(A)$ is a reflexive Azumaya algebra and by Lemma VI.3.13 it is also an Azumaya algebra. Consequently $Q_p(A)$ is an Azumaya algebra too, and this holds at every $p \in \mathrm{Proj}(R)$. ∎

VI.3.15 Remark This provides an easy proof and also a generalization of Theorem 3.19 in [105]. In fact, the Brauer group of a ringed space is defined in terms of locally separable sheaves of $\tilde{R}$-algebras where $\tilde{R}$ is the structure sheaf on Spec(R). For such sheaves the stalks are Azumaya algebras over the stalks $\tilde{R}_p, p \in \mathrm{Spec}(R)$. Similarly for a locally separable sheaf of graded $\tilde{R}$-algebras where now $\tilde{R}$ is the graded structure sheaf over Proj(R) (note: not the structure sheaf usually defined by going to degree zero everywhere locally): each $Q_p^g(A) = \tilde{A}_p$ for $p \in \mathrm{Proj}(R)$ is an Azumaya algebra. In this terminology, Theorem VI.3.14 shows that, if $\dim(R) \leq 3$, the graded structure sheaf of a reflexive graded Azumaya algebra A corresponds to an element of $\mathrm{Br}(\mathrm{Proj}(R))$. Obviously, a graded R-algebra A such that for all $p \in \mathrm{Proj}(R)$, $Q_p^g(A)$ is an Azumaya algebra, will always be a reflexive Azumaya algebra, because $X_g^{(1)}(R) \subset \mathrm{Proj}(R)$ unless $R = k[X]$, a case we can exclude for obvious reasons. That this correspondence does define an isomorphism $\mathrm{Br}(\mathrm{Proj}(R)) \cong \beta_g(R)$ is easily verified by checking that the equivalence relations introduced in the definition of both groups correspond. So for a connected regular projective variety over a field k, X say, of dimension at most two, we do recover the result $\mathrm{Br}(X) = \beta_g(\Gamma_*(X))$.

Let us now study the singular case when X is a curve. First we summarize some easy results concerning the graded conductor ideal for integral closures of graded domains. Let S be a commutative graded domain and $\bar{S}$ the integral closure of S in its field of fractions; the conductor of $\bar{S}$ in S is defined to be $c = \{x \in \bar{S} : rx \in S \text{ for all } r \in \bar{S}\}$. Obviously c is a graded ideal of S and of $\bar{S}$.

VI.3.16 Lemma Let R be a positively graded noetherian domain and let c be the conductor of $\bar{R}$ in R. For any homogeneous f in R_+ we have:

(3.16.1) $Q_f^g(\bar{R})$ is the integral closure of $Q_f^g(R)$.
(3.16.2) The conductor d of $Q_f^g(\bar{R})$ in $Q_f^g(R)$ is exactly $Q_f^g(c)$.

Proof: (3.16.1) is obvious because $Q_f^g(R)$ is just localization at the homogeneous multiplicative set $\{1, f, f^2, \ldots\}$.
(3.16.2) Clearly $Q_f^g(c) \subset d$. Conversely, since $\bar{R}$ is a finitely generated R-module, say $\bar{R} = Rx_1 + \cdots + Rx_n$, there is for each element $y \in d$ some $m \in \mathbf{N}$ large enough so that $f^m y x_i \in R$ for $i = 1, \ldots, n$. Consequently, $f^m y \bar{R} \subset R$ and thus $f^m y \in c$, that is, $y \in Q_f^g(c)$. ∎

Using the local description of the normalization of a scheme in terms of an affine covering, we obtain:

VI.3.17 Corollary Let R be a positively graded noetherian domain; then $\mathrm{Proj}(\bar{R})$ is the normalization of $\mathrm{Proj}(R)$. If R' is a graded ring containing R and integral over R, then the inclusion $R \subset R'$ induces a scheme morphism $\mathrm{Proj}(R') \to \mathrm{Proj}(R)$. In particular, $\mathrm{Proj}(\bar{R}) \to \mathrm{Proj}(R)$.

Now fix a field k and assume that R is a positively graded k-algebra with $R_0 = k$, such that R is generated as a k-algebra by a finite number of homogeneous elements in R_1. Let $\bar{R}$ be the integral closure of R and c the conductor of $\bar{R}$ in R. Assume that $X = \mathrm{Proj}(R)$ is a curve; then $X = \mathrm{Proj}(\bar{R})$ is its normalization and c its conductor sheaf $\mathrm{Ann}_X(\pi_* \underline{O}_{\tilde{X}} / \underline{O}_X)$, where $\pi : \tilde{X} \to X$ is the canonical covering. Let V (resp. $\tilde{V}$) denote the closed subscheme of X (resp. $\tilde{X}$) determined by the ideal c. With this notation:

VI.3.18 Lemma $\mathrm{Pic}(\tilde{V}) = 0$.

Proof: By definition, $\mathrm{Pic}(\tilde{V}) = H^1(\tilde{V}, \underline{0}^*_{\tilde{V}})$ and $\dim \tilde{V} = 0$ because $\tilde{V} \cong \mathrm{Proj}(\bar{R}/c)$ and $\bar{R}/c$ is gr-semilocal. It follows that $H^1(\tilde{V}, \underline{0}^*_{\tilde{V}}) = 0$ (cf. [93]). ■

VI.3.19 Theorem There is an exact sequence of abelian groups

$$1 \longrightarrow \mathrm{Br}(X) \longrightarrow \mathrm{Br}(\tilde{X}) \oplus \mathrm{Br}(V) \longrightarrow \mathrm{Br}(\tilde{V})$$

Proof: X may be covered by two open affine schemes having an open affine as their intersection because X is a separated scheme. The same statement holds for all schemes appearing in the following diagram:

$$\begin{array}{ccc} \tilde{X} & \xrightarrow{i_1} & \tilde{V} \\ \downarrow{\scriptstyle \pi} & & \downarrow \\ X & \xrightarrow{i} & V \end{array}$$

From [31], we recall that there is an exact sequence, in the étale topology, of sheaves on X:

$$0 \longrightarrow G_{m,X} \longrightarrow \pi_* G_{m,\tilde{X}} \oplus i_* G_{m,V} \longrightarrow (\pi i_1)_* G_{m,\tilde{V}} \longrightarrow 0$$

that yields a long exact sequence (we denoted $G_m = U$):

$$\cdots \longrightarrow \mathrm{Pic}(\tilde{V}) \longrightarrow H^2(X_{\text{ét}}, U) \longrightarrow H^2(\tilde{X}_{\text{ét}}, U) \oplus H^2(V_{\text{ét}}, U) \longrightarrow H^2(\tilde{V}_{\text{ét}}, G_m) \longrightarrow \cdots$$

Using that $\mathrm{Pic}(\tilde{V}) = 0$ (see the lemma) and Gabber's theorem (cf. [26, 31]) we obtain the desired exact sequence. ■

In view of Remark VI.3.15, we have $\mathrm{Br}(X) = \beta_g(\Gamma_*(X))$, and we usually consider $R = \Gamma_*(X)$, for example, when R is integrally closed. In view of Theorem VI.3.8 we obtain $\mathrm{Br}(X) \cong \mathrm{Br}_g(R)$ for a regular curve. We now focus on the singular case. Note that V contains only finitely many graded prime ideals, say $V = \{p_1, \ldots, p_n\}$ and we may view $p_i, i = 1, \ldots, n$, as an element of the closed space defined by the conductor ideal c in $\mathrm{Proj}(R)$ or as an element of $\mathrm{Proj}(R/c)$. Since $\bigcap_{i \neq j} p_j \not\subset p_i$ we may select $f_1, \ldots, f_n \in h(R_+)$ such that the open set induced in V by f_i consists exactly of p_i. Therefore V is the disjoint union of the affines $\mathrm{Spec}(Q^g_{f_i}(R/c)_0) = \mathrm{Spec}((Q^g_{p_i}(R)/Q^g_{p_i}(c))_0) = \mathrm{Spec}(B_i)$, hence $\mathrm{Br}(V) = \oplus_{i=1}^n \mathrm{Br}(B_i)$. It is easily verified that $B_i/\mathrm{rad}(B_i)$ is nothing but the residue field $K_X(p_i)$ (noting that $\mathrm{rad}(Q^g_{p_i}(c)) = Q^g_{p_i}(p_i)$ and $((Q^g_{p_i}(R)/Q^g_{p_i}(c))_0)_{\mathrm{red}} = K_X(p_i))$. Taking into account that $\mathrm{Br}(S) = \mathrm{Br}(S_{\mathrm{red}})$ holds generally, we may combine the foregoing remarks into the following:

VI.3.20 Theorem Let $X = \mathrm{Proj}(R)$ be a connected projective curve, let $\bar{R}$ be the integral closure of R, and c the conductor of $\bar{R}$ in R. Let V (resp. $\tilde{V}$) be the closed subschemes of X (resp. $\tilde{X} = \mathrm{Proj}(\bar{R})$) determined by c. Then there is an exact sequence of abelian groups

$$0 \longrightarrow \mathrm{Br}(X) \longrightarrow \beta_g(R) \oplus (\oplus_{p \in V} \mathrm{Br}(K_X(p))) \longrightarrow \oplus_{q \in \tilde{V}} \mathrm{Br}(K_{\tilde{X}}(q)).$$

For higher dimensional schemes (varieties) it is necessary to replace $\beta_g(R)$ in the foregoing theorem by a relative Brauer group $\mathrm{Br}_g(R, \kappa_+)$ associated to the kernel functor κ_+ defined by the Gabriel topology

$$\mathcal{L}(\kappa_+) = \{L \text{ideal of } R : L \supset R_+^n \text{for some } n \in \mathbf{N}\}$$

Under some "regularity in codimension m" condition one might hope to replace $\mathrm{Br}_g^{(1)}(R) = \beta_g(R)$ by $\mathrm{Br}_g^{(m)}(R)$ and repeat most of the foregoing theory. There are several problems here, for example: Can one have an analog for Theorem VI.3.8 under the "regularity in codimension m" condition? Is then $\mathrm{Br}^{(m)}(R) = \beta(R)$? Is there a version of Theorem VI.3.10 in the case where R is regular in codimension m? Is $\mathrm{Br}(\mathrm{Proj}(R))$ determined by $\mathrm{Br}_g^{(m)}(\Gamma_*(R))$ in that case? Other, perhaps interesting, problems remain in deriving suitable graded versions of Theorem VI.3.10, using the graded étale cohomology.

References

1. D. D. Anderson, D. F. Anderson, Divisorial ideals and invertible ideals in a graded integral domain, *J. Algebra* 76 (1982), 549–569.

2. M. Artin, On the joins of Hensel rings, *Adv. in Math.* 7 (1971), 282–296.

3. M. Artin, B. Mazur, Étale homotopy, *Lecture Notes in Math. 100*, Springer-Verlag, Berlin, 1969.

4. M. Artin, A. Grothendieck, J.-L. Verdier, Théorie des topos et cohomologie étale des schémas, *Lecture Notes in Math. 269*, *270*, *305*, Springer-Verlag, Berlin, 1972–1973.

5. M. Auslander, O. Goldman, The Brauer group of a commutative ring, *Trans. Amer. Math. Soc.* 97 (1960), 367–409.

6. H. Bass, *Algebraic K-Theory*, Benjamin, New York, 1968.

7. H. Bass, *Lectures on Topics in Algebraic K-Theory*, Tata Institute for fundamental research, Bombay, 1967.

8. I. Bucur, A. Deleanu, *Introduction to the Theory of Categories and Functors*, Wiley, London, 1968,

9. S. Caenepeel, A cohomological interpretation of the graded Brauer group I, *Comm. Algebra* 11 (1983), 2129–2149.

10. S. Caenepeel, A cohomological interpretation of the graded Brauer group II, *J. Pure Appl. Algebra* 38 (1985), 19–38.

11. S. Caenepeel, Gr-complete and Gr-Henselian Rings, in *Methods in Ring Theory*, Reidel, Dordrecht, The Netherlands, 1984.

12. S. Caenepeel, On Brauer groups of graded krull domains and positively graded rings, *J. Algebra* 99 (1986), 466—474.

13. S. Caenepeel, A Graded Version of Artin's Refinement Theorem, in Ring theory 1985, *Lecture Notes in Math. 1197*, Springer-Verlag, Berlin, 1986.

14. S. Caenepeel, A cohomological description of graded and relative Brauer groups, thesis, Free University of Brussels, VUB, 1984.

15. S. Caenepeel, M. Van den Bergh, F. Van Oystaeyen, Generalized crossed products applied to Maximal orders, Brauer groups and related exact sequences, *J. Pure Appl. Algebra* 33 (1984), 123–149.

16. S. Caenepeel, F. Van Oystaeyen, Crossed products over graded local rings, in Brauer groups in ring theory and algebraic geometry, *Lecture Notes in Math. 917*, Springer-Verlag, Berlin, 1982.

17. H. Cartan, S. Eilenberg, *Homological Algebra*, Princeton University Press, Princeton, 1956.

18. S. Chase, P. Harrison, A. Rosenberg, Galois Theory and cohomology of commutative rings, *Mem. Am. Math. Soc.* 52 (1965), 1–19.

19. S. Chase, A. Rosenberg, Amitsur cohomology and the Brauer group, *Mem. Am. Math. Soc.* 52 (1965), 20–45.

20. E. Dade, Compounding Clifford's theory, *Ann. of Math.* 2, 91 (1970), 236–270.

21. F. DeMeyer, The Brauer group of some separably closed rings, *Osaka J. Math. 3* (1966), 201–204.

22. F. DeMeyer, The Brauer group of a ring modulo an ideal, *Rocky Mountain J. Math.* 6 (1976), 191–198.

23. F. DeMeyer, E. Ingraham, Separable algebras over commutative rings, *Lecture Notes in Math. 181*, Springer-Verlag, Berlin, 1971.

24. T. Ford, On the Brauer group of a Laurent polynomial ring, to appear

25. R. Fossum, *The Divisor Class Group of a Krull Domain*, Ergebnisse der Mathematik 74, Springer-Verlag, Berlin, 1973.

26. O. Gabber, Some Theorems on Azumaya Algebras, in Groupe de Brauer, *Lecture Notes in Math. 844*, Springer-Verlag, Berlin, 1981.

27. J. Giraud, *Cohomologie Non Abélienne*, Springer-Verlag, Berlin, 1971.

28. A. Grothendieck, Le Groupe de Brauer I, II, III, in *Dix Exposés sur la Cohomologie des Schémas*, North Holland, Amsterdam, 1968.

29. I. N. Herstein, Noncommutative Rings, *Carus Math. Monographs 15*, Math. Ass. of America, Chicago, 1968.

30. K. Hirata, Some types of separable extensions of rings, *Nagoya Math. J.* 33 (1968), 108–115.

31. R. Hoobler, When is $\mathrm{Br}(X) = \mathrm{Br}'(X)$?, in Brauer Groups in Ring Theory and Algebraic Geometry, *Lecture Notes in Math. 917*, Springer-Verlag, Berlin, 1982.

32. R. Hoobler, Functors of Graded Rings, in *Methods in Ring Theory*, Reidel, Dordrecht, The Netherlands, 1984.

33. E. Ingraham, On the existence and conjugacy of inertial subalgebras, *J. Algebra* 31 (1974), 547–556.

34. T. Kanzaki, On generalized crossed products and Brauer groups, *Osaka J. Math.* 5 (1968), 175–188.

35. I. Kaplansky, *Commutative Rings*, University of Chicago Press, Chicago, 1974.

36. M. A. Knus, A teichmüller cocycle for finite extensions, Preprint.

37. M. A. Knus, M. Ojanguren, Théorie de la descente et algébres d'Azumaya, *Lectures Notes in Math. 389*, Springer-Verlag, Berlin, 1974.

38. M. A. Knus, M. Ojanguren, A Mayer-Vietoris sequence for the Brauer group, *J. Pure Appl. Algebra* 5 (1974), 345–360.

39. M. A. Knus, M. Ojanguren, Cohomologie étale et groupe de Brauer, in Groupe de Brauer, *Lecture Notes in Math. 844*, Springer-Verlag, Berlin, 1981.

40. T. Lam, Serre's conjecture, *Lecture Notes in Math. 635*, Springer-Verlag, Berlin, 1978.

41. L. Le Bruyn, F. Van Oystaeyen, Generalized Rees rings and relative maximal orders satisfying polynomial identities, *J. Algebra* 83 (1983), 409–436.

42. L. Le Bruyn, M. Van den Bergh, F. Van Oystaeyen, *Graded Orders*, Birkhäuser Verlag, Boston, 1988.

43. R. Matsuda, On algebraic properties of infinite group rings, *Bull. Fac. Sci. Ibaraki Univ. Series A. Math.* 7 (1975), 29–37.

44. J. S. Milne, *Étale Cohomology*, Princeton University Press, Princeton, 1980.

45. Y. Miyashita, An Exact sequence associated with a generalized crossed product, *Nagoya Math. J.* 49 (1973), 21-51.

46. M. Nagata, Local rings, *Inter Science Tracts in Pure and Appl. Math. 13*, New York, 1962.

47. C. Năstăsescu, E. Nauwelaerts, F. Van Oystaeyen, Arithmetically graded rings revisited, *Comm. Algebra* 14 (1986), 1991–2008.

48. C. Năstăsescu, F. Van Oystaeyen, Graded and filtered rings and modules, *Lecture Notes in Math. 758*, Springer-Verlag, Berlin, 1980.

49. C. Năstăsescu, F. Van Oystaeyen, *Graded Ring Theory*, *Library of Math. 28*, North Holland, Amsterdam, 1982.

50. C. Năstăsescu, F. Van Oystaeyen, On strongly graded rings and crossed products, *Comm. Algebra* 10 (1982), 2085–2106.

51. D. Northcott, *Homological Algebra*, Cambridge University Press, Cambridge, 1960.

52. M. Ojanguren, R. Sridharan, forthcoming publication.

53. M. Orzech, C. Small, *The Brauer Group of a Commutative Ring*, Marcel Dekker, Inc., New York, 1975.

54. M. Raynaud, Anneaux locaux Henséliens, *Lecture Notes in Math. 169*, Springer-Verlag, Berlin, 1971.

55. J. P. Serre, Cohomologie Galoisienne, *Lecture Notes in Math. 5*, Springer-Verlag, Berlin, 1964.

56. J. P. Serre, Algébre locale—multiplicités, *Lecture Notes in Math. 11*, Springer-Verlag, Berlin, 1965.

57. M. Van den Bergh, Graded dedekind rings, *J. Pure and Appl. Algebra* 35 (1985), 105–115.

58. M. Van den Bergh, A note on graded Brauer groups, *Bull. Soc. Math. Belg.* 39 (1987), 177–179.

59. M. Van den Bergh, A note on graded K-theory, *Comm. Algebra* 14 (1986), 1561–1564.

60. J. P. Van Deuren, F. Van Oystaeyen, Arithmetically graded rings, in Ring Theory, Antwerp 1980, *Lecture Notes in Math. 825*, Springer-Verlag, Berlin, 1981.

61. J. Van Geel, F. Van Oystaeyen, About graded fields, *Indag. Math.* 43 (1981), 273–286.

62. F. Van Oystaeyen, On Brauer groups of arithmetically graded rings, *Comm. Algebra* 9 (1981), 1873–1892.

63. F. Van Oystaeyen, Crossed products over arithmetically graded rings, *J. Algebra* 80 (1983), 537–551.

64. F. Van Oystaeyen, Graded Azumaya algebras and Brauer groups, in Ring Theory, Antwerp 1980, *Lecture Notes in Math. 825*, Springer-Verlag, Berlin, 1981.

65. F. Van Oystaeyen, Some constructions of rings, *J. Pure Appl. Algebra* 31 (1984), 241–252.

66. F. Van Oystaeyen, Generalized Rees rings and arithmetically graded rings, *J. Algebra* 82 (1983), 185–193.

67. F. Van Oystaeyen, On Clifford systems and generalized crossed products, *J. Algebra* 87 (1984), 396–415.

68. F. Van Oystaeyen, Azumaya strongly graded rings and Ray classes, *J. Algebra* 103 (1986), 228–240.

69. F. Van Oystaeyen, A. Verschoren, *Relative Invariants of Rings Part I and II*, Marcel Dekker, Inc., New York, 1983 and 1984.

70. A. Van Rooij, *Non-Archimedean Functional Analysis*, Marcel Dekker, Inc., New York, 1978.

71. A. Verschoren, Mayer-Vietoris sequences for Brauer groups of graded rings, *Comm. Algebra* 10 (1982), 765–782.

72. O. E. Villamayor, D. Zelinsky, Brauer groups and Amitsur cohomology for general commutative ring extensions, *J. Pure Appl. Algebra* 10 (1977), 19–55.

73. C. Weibel, Homotopy in algebraic K-theory, thesis, University of Chicago, 1977.

74. D. Zelinsky, Long exact sequences and the Brauer group, in Brauer Groups, Evanston 1975, *Lecture Notes in Math. 549*, Springer-Verlag, Berlin, 1976.

75. D. Zelinsky, Brauer Groups, in *Ring Theory II*, Marcel Dekker, Inc., New York, 1977.

76. M. Auslander, O. Goldman, Maximal orders, *Trans. Amer. Math. Soc.* 97 (1960), 1–24.

77. M. Beattie, A direct sum decomposition for the Brauer group of H-module algebras, *J. Algebra* 43 (1976), 686–693.

78. M. Beattie, The Brauer group of central separable G-Azumaya algebras, *J. Algebra* 54 (1978), 516–525.

79. M. Beattie, Automorphisms of G-Azumaya algebras, *Canad. J. Math.* 37 (1985), 1047–1058.

80. M. Beattie, The subgroup structure of the Brauer group of RG-dimodule algebras, in Ring Theory 1985, *Lecture Notes in Math. 1197*, Springer-Verlag, Berlin, 1987.

81. M. Beattie, Computing the Brauer group of graded Azumaya algebras from its subgroups, *J. Algebra* 101 (1986), 339–349.

82. M. Beattie, S. Caenepeel, The Brauer–Long group of $\mathbf{Z}/p^r\mathbf{Z}$-dimodule algebras, in preparation.

83. S. Caenepeel, Cancellation theorems for projective graded modules, in Ring Theory, Granada 1986, *Lecture Notes in Math. 1324*, Springer-Verlag, Berlin 1988.

84. S. Caenepeel, A. Verschoren, A relative version of the Chase-Harrison-Rosenberg Sequence, *J. Pure Appl. Algebra* 41 (1986), 149–168.

85. S. U. Chase, M. E. Sweedler, Hopf algebras and Galois Theory, *Lecture Notes in Math. 97*, Springer-Verlag, Berlin, 1969.

86. L. N. Childs, The Brauer group of graded Azumaya algebras II: Graded Galois extensions, *Trans. Amer. Math. Soc.* 204 (1975), 137–160.

87. L. N. Childs, G. Garfinkel, M. Orzech, The Brauer group of graded Azumaya algebras, *Trans. Amer. Math. Soc.* 175 (1973), 299–326.

88. L. Claborn, R. Fossum, Generalization of the notion of class group, *Ill. J. Math.* 12 (1968), 228–253.

89. A. P. Deegan, A subgroup of the generalized Brauer group of G-Azumaya algebras, *J. London Math. Soc.* 23 (1981), 223–240.

90. F. DeMeyer, T. Ford, Computing the Brauer-Long group of $\mathbf{Z}/2$-dimodule algebras, *J. Pure Appl. Algebra*, to appear.

91. A. Fröhlich, C. T. C. Wall, Generalizations of the Brauer group I, preprint.

92. D. K. Harrison, Abelian extensions of commutative rings, *Mem. Amer. Math. Soc.* 52 (1965), 66–79.

93. R. Hartshorne, Algebraic geometry, *Graduate Texts in Math. 52*, Springer-Verlag, Berlin, 1977.

94. M. A. Knus, Algebras graded by a group, in Category theory, homology theory and their applications II," *Lecture Notes in Math. 92*, Springer-Verlag, Berlin, 1969.

95. L. Le Bruyn, A cohomological interpretation of the reflexive Brauer group, *J. Algebra* 105 (1987), 250–254.

96. F. W. Long, A generalization of the Brauer group of graded algebras, *Proc. London Math. Soc.* 29 (1974), 237–256.

97. F. W. Long, The Brauer group of dimodule algebras, *J. Algebra* 30 (1974), 559–601.

98. M. Orzech, On the Brauer group of modules having a grading and an action, *Canad. J. Math.* 28 (1976), 533–552.

99. M. Orzech, Correction to On the Brauer group of modules having a grading and an action, *Can. J. of Math.* 32 (1980), 1523–1524.

100. D. S. Rim, An exact sequence in Galois cohomology, *Proc. Amer. Math. Soc.* 16 (1965), 837–840.

101. C. Small, The Brauer-Wall group of a commutative ring, *Trans. Amer. Math. Soc.* 156 (1971), 455–491.

102. C. Small, The group of quadratic extensions, *J. Pure Appl. Algebra* 2 (1972), 83–105.

103. F. Tilborghs, The Brauer group of R-algebras which have compatible G-action and $\mathbf{Z} \times G$-grading, preprint.

104. F. Tilborghs, F. Van Oystaeyen,. Brauer-Wall algebras graded by $\mathbf{Z}_2 \times \mathbf{Z}$, *Comm. Algebra*, to appear.

105. F. Van Oystaeyen, A. Verschoren, On the Brauer group of a projective variety, *Israel J. Math.* 42 (1982), 37–59.

106. F. Van Oystaeyen, A. Verschoren, On the Brauer group of a projective curve, *Israel J. Math.* 42 (1982), 327–333.

107. S. Yuan, Reflexive modules and algebras class groups over Noetherian integrally closed domains, *J. Algebra* 32 (1974), 405–417.

108. C. T. C. Wall, Graded Brauer groups, *J. Reine Angew. Math.* 213 (1964), 187–199.

Index

Acyclic object, 158
Artin's refinement theorem, 75

Bass' exact sequence, 67
Brauer–Long group, 204
Brauer–Wall group, 205, 228

Chase–Rosenberg exact sequence, 15, 120, 123, 184, 218
Clifford system, 17
Coskeleton functor, 169
Crossed product theorems, 129

Discrete gr-valuation, 27
Divisorially graded ring, 23

Embedding theorem, 45
Etale site, 162

Flat site, 162

Gabber's theorem, 188, 219, 248
Generalized crossed product, 11
Genus, 33
G-Azumaya algebra, 203

G-dimodule, 202
G-dimodule algebra, 202
G-module, 202
G-module algebra, 202
Graded absolutely integrally closed ring, 61
Graded approximation property, 32
Graded Azumaya algebra, 105
Graded Brauer group, 110
Graded Brauer group of Childs, Garfinkel, and Orzech, 205
Graded completion, 48
Graded étale algebra, 40
Graded field, 2
Graded Galois extension, 44
Graded isomorphic, 2
Graded Jacobson radical, 2
Graded module, 1
Graded morphism, 2
Graded nilradical, 2
Graded Picard group, 91
Graded residue class field, 35
Graded ring of type G, 1
Graded separable algebra, 40
gr-acyclic ring, 71
gr-algebraically closed graded field, 46
gr-central simple algebra, 154
gr-decomposed algebra, 54
gr-Dedekind ring, 29
gr-field, 2
gr-henselian ring, 54
gr-I-complete, 47
gr-local ring, 3
gr-maximal ideal, 2
gr-maximal order, 155
gr-prime ideal, 2
gr-principal ideal domain, 29
gr-projective module, 83
gr-quasiacyclic ring, 72
gr-quasifinite algebra, 36
gr-separably closed graded field, 46
gr-simple algebra, 151
gr-valuation ring, 26

Homogeneous element, 1
Hypercovering, 169

I-adically complete ring, 47
I-adic topology, 47
Invertible module, 3

Krull's principal ideal theorem, 153

Leray spectral sequence, 160
Locally homogeneous element or morphism, 202

Presheaf, 162
Principal gr-valuation, 27

Quadratic extension, 229
Quasistrongly graded ring, 72

Right derived functor, 157
Rosenberg–Zelinsky exact sequence, 223

Sheaf, 162
Serre's theorem, 88
Smash product, 212
Standard gr-étale algebra, 40
Strict gr-henselization, 58
Strictly gr-henselian ring, 58
Strongly graded ring, 1

Unimodular element, 86

Verdier's refinement theorem, 169

Zariski site, 162
Zariski's main theorem, 36